Serious Managers Guide to AI with FHIR

AF894245

Claude Louis-Charles, Phd

Cybersoft Publishing LLC

Fort Washington, MD 20744

Drclaude.net

First Edition April 2026

Contents

Introduction

Most healthcare managers did not sign up to become experts in data standards or artificial intelligence. You signed up to deliver care, keep operations running, meet regulatory obligations, and protect your teams from being overwhelmed. Yet, almost every major challenge on your plate today—fragmented systems, rising expectations, workforce strain, and tightening margins—now has "AI" and "interoperability" attached to it. You are expected to make smart decisions about technologies you didn't ask for, in an environment that rarely slows down long enough to explain them.

This book is tailored to address real-world challenges. It presents **FHIR** (Fast Healthcare Interoperability Resources) and **AI** as practical tools, not just technical concepts, aimed at solving everyday problems: missing information where it's needed most, staff overwhelmed by repetitive tasks, patients navigating fragmented care experiences, and leaders questioning the value of digital investments. Instead of programming code, you'll find straightforward explanations, relevant examples, and frameworks geared toward management—helping you consider whether an approach makes sense, how to implement it, what risks might arise, and how to measure success.

At the heart of the book is a simple idea: FHIR and AI are two halves of a modern healthcare engine. FHIR provides a standard way to exchange and structure data across systems. AI gives you a standard set of patterns to turn that data into predictions, priorities, classifications, and drafts that change how work gets done. When you connect them deliberately—through governance, secure infrastructure, and well-designed workflows—you get something more than isolated projects. You get the beginnings of a learning health system: an organization that gets a bit smarter, safer, and more efficient with every encounter.

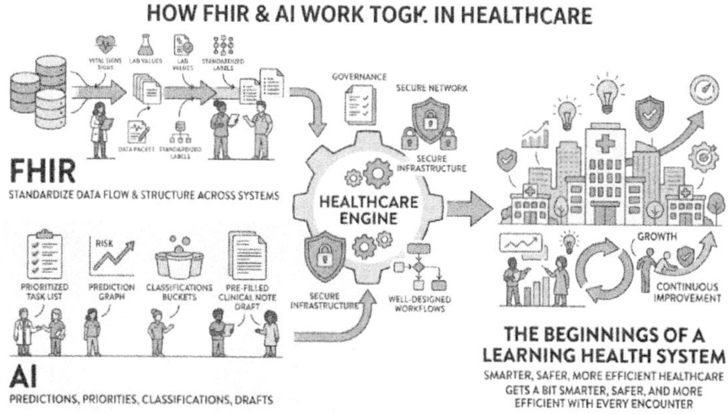

The material guides you from strategy to execution. The initial chapters clarify the importance of FHIR and AI and introduce their basics. The middle sections cover designing integrated workflows and establishing early governance and responsible AI practices. Later chapters address scaling up, building an AI-ready FHIR backbone, progressing to production, and adopting evidence-driven processes. The emphasis remains on decisions, roles, and workflows rather than technology alone.

This is not a book about chasing the latest algorithm or buying a magic platform. It is a book about building **operational clarity** in a noisy landscape so that you can say “yes” to the right AI and interoperability ideas and “not yet” or “not like this” to the rest. If you are accountable for outcomes in healthcare—clinical, operational, financial, or regulatory—this is meant to sit on your desk as a working guide. You will not become a FHIR engineer or a machine-learning researcher by reading it. You will become the person in the room who can connect technical proposals back to mission, value, feasibility, and risk—and who can lead your teams through the FHIR + AI era with confidence.

1 The New Healthcare Partnership: FHIR + AI

On a Tuesday morning in a busy urban hospital, a care coordinator opens her dashboard. Instead of the usual chaos—scattered spreadsheets, three different systems for patient histories, and a flood of unfiltered alerts—she sees something different. A clean, prioritized list of patients most at risk of readmission appears on her screen. Each patient's record shows relevant clinical data pulled automatically from the EHR, recent lab results, active medications, and social determinants that matter—all in one place, in a format she can understand immediately.

Behind the scenes, FHIR (Fast Healthcare Interoperability Resources) is quietly moving standardized data between systems. AI analyzes that data to surface patterns that would take her hours to find manually. The system flags Mr. Thompson, a 68-year-old with worsening heart failure markers and three missed follow-up appointments. It suggests prioritizing outreach before his condition escalates. She reviews the recommendation, agrees with the assessment, calls Mr. Thompson, and schedules an urgent visit. Readmission is avoided.

This is not science fiction. This is the new healthcare partnership: FHIR enabling clean, consistent data flow, and AI turning that data into actionable intelligence, with humans firmly in command of the decisions that matter.

For managers, this partnership is not primarily a technology story. It is an operational one. It changes how work gets done, how decisions are made, and how your organization learns and improves. This chapter introduces that partnership and why it matters to you right now.

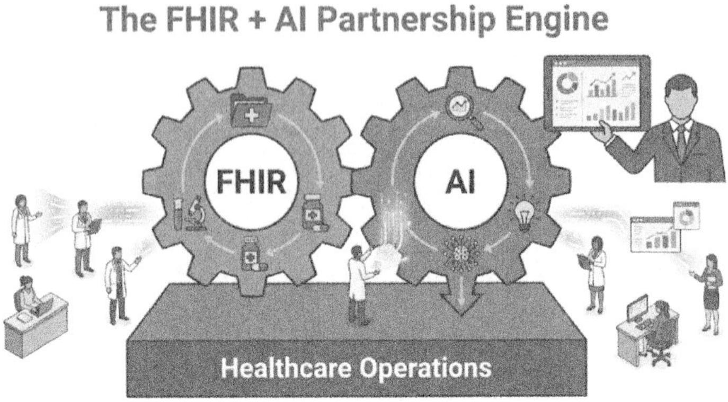

What This Partnership Actually Means

FHIR and AI are often discussed separately—FHIR in interoperability meetings, AI in innovation summits. That separation misses the point. Neither is transformative on its own. Together, they form a modern healthcare engine.

FHIR is your data backbone. It standardizes how patient information—demographics, encounters, diagnoses, medications, lab results, care plans—is represented and exchanged across systems. Instead of custom integrations for every vendor and every use case, you have a common language. That common language makes data more accessible, governable, and reusable.

AI is your intelligence layer. It detects patterns in data that humans cannot see at scale, predicts outcomes before they happen, prioritizes work so scarce resources go where they matter most, and automates repetitive tasks so people can focus on judgment and relationships. But AI only works well when it has clean, structured, consistent data to learn from. That is where FHIR comes in.

The partnership works like this:

- FHIR feeds AI. Standardized resources (Patient, Encounter, Observation, Condition, MedicationRequest) become the training data and operational inputs for AI models. Because FHIR is consistent, AI can learn patterns across sites, populations, and time periods without constantly reengineering pipelines.
- AI enriches FHIR workflows. AI-derived insights—risk scores, care gaps, predicted costs, recommended interventions—flow back into FHIR-enabled systems as Observations, Flags, or extensions. Clinicians, care managers, and operational staff see intelligence embedded in their work.
- Humans decide and act. FHIR and AI prepare the ground, but humans make the calls. The care coordinator reviewing Mr. Thompson's case decides whether to call, what to say, and how urgent the situation is. The system supports the human leads.

This loop—data flows, intelligence operates, humans act, outcomes feed back—is the fundamental pattern you will see throughout this book.

Why Managers Should Care Now

You might be thinking: "We are already overwhelmed with EHR implementations, regulatory mandates, and workforce shortages. Why add FHIR and AI to the pile?"

The answer is that FHIR and AI are not add-ons. They are how you address the problems already on your desk.

Healthcare organizations face relentless pressure:

- **Administrative waste** – Staff spends hours hunting for information across disconnected systems, manually reconciling data, and chasing down missing documentation.
- **Workforce strain** – Clinicians burn out from documentation burdens and low-value tasks. Care managers juggle impossibly large panels. Analysts drown in report requests.
- **Fragmented data** – Patient information is scattered across EHRs, claims systems, labs, pharmacies, and external partners. No one has the full picture.
- **Regulatory demands** – Interoperability mandates (TEFCA, information blocking rules, payer data exchange requirements) are not optional. Compliance is table stakes.
- **Rising expectations** – Patients expect seamless, coordinated care. Payers demand performance on value-based contracts. Boards want evidence that digital investments are paying off.

FHIR addresses the data problem. It gives you a structured way to break down information silos and move data where it needs to go—internally and with external partners. AI addresses the intelligence problem. It helps you do more with less: prioritize better, predict sooner, and automate intelligently.

Together, they create new operational capacity. They free up time, reduce errors, surface insights, and enable workflows that were previously impossible at scale. For managers, this is not about chasing innovation for its own sake. It is about building a foundation that lets your organization deliver better outcomes with the resources you actually have.

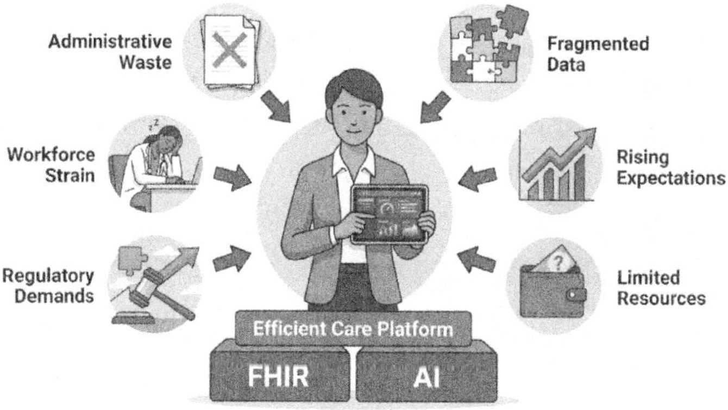

The Symbiosis Model: How FHIR and AI Work Together

To understand the partnership at a practical level, think of it as a symbiotic loop with four stages: **data flow, insight flow, action flow, and learning flow**.

Stage 1: Data Flow (FHIR)

Clean, standardized patient data moves from source systems—EHRs, labs, claims, devices—into a FHIR-enabled environment. This might be a FHIR server, a data lake with FHIR APIs, or a cloud-based interoperability platform. The key is that data is represented using FHIR resources and accessible via FHIR APIs, not locked in proprietary formats.

At this stage, you are establishing the backbone. Which FHIR resources matter most for your use cases? What quality and completeness standards must be met? How will you handle data from multiple sources that do not perfectly align? These are governance and architecture questions, not just technical ones.

Stage 2: Insight Flow (AI)

AI models consume FHIR data and generate insights. A readmission risk model reads Encounter, Condition, Observation, and MedicationRequest resources to predict which patients are at high risk of readmission. A care gap identification tool scans Observation and Procedure resources to find patients overdue for screenings. A documentation assistant uses recent Encounter and Condition data to draft clinical notes.

The insights produced—risk scores, recommendations, summaries, predictions—are structured and traceable. Ideally, they are also represented in FHIR (as Observation resources, RiskAssessment resources, or extensions) so they can flow back into the operational environment.

Stage 3: Action Flow (Humans)

Insights reach the people who can act on them: clinicians, care managers, operational staff, and administrators. They see AI outputs embedded in familiar tools—dashboards, EHR screens, work queues, alerts. Crucially, they retain control. They can review the underlying data (thanks to FHIR's transparency), question the recommendation, override it, or escalate for help.

The care coordinator who prioritizes Mr. Thompson is not mindlessly following the AI. She reviews his recent vitals, checks his appointment history, considers the social factors the model flagged, and applies her professional judgment. The AI prepared the case; she made the decision.

Stage 4: Learning Flow (Feedback)

Outcomes are documented—did the outreach happen? Did the patient's condition improve? Was the prediction accurate?—and fed back into the FHIR environment. Over time, this feedback improves the AI models. It also improves workflows, policies, and training. The loop becomes a continuous learning system.

This four-stage loop is not theoretical. It is the operational pattern behind successful FHIR + AI implementations. When any stage is weak—dirty data, opaque models, poor integration, no feedback—the loop breaks and value disappears.

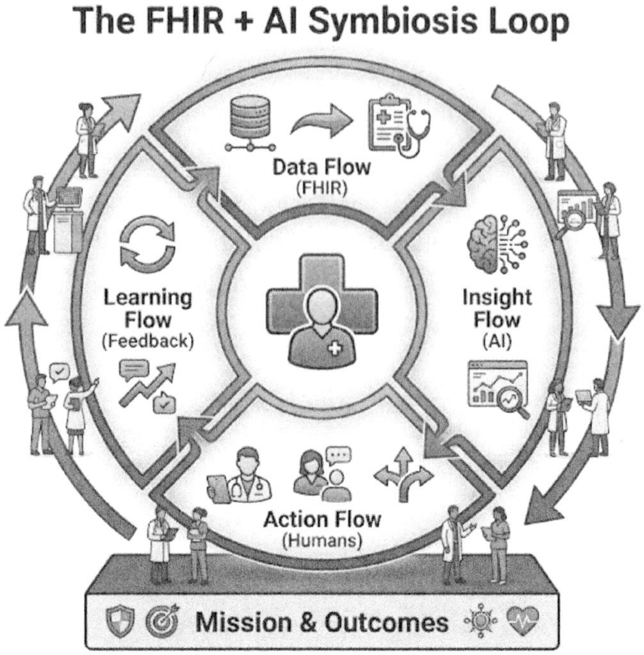

Real-World Examples of the Partnership in Action

Abstract models are useful, but concrete examples make the partnership tangible. Here are three scenarios in which FHIR and AI work together to change how healthcare is delivered.

Example 1: Care Coordination at Scale

A health system serves 80,000 patients with chronic conditions. Care managers cannot possibly touch every patient every month. Instead, FHIR-enabled data from the EHR, claims, and health information exchanges flows into a care management platform. An AI model analyzes recent encounters, lab trends, medication adherence, and social risk factors. It produces a daily prioritized list of patients who need outreach—those whose conditions are worsening, who missed appointments, or who have new risk factors.

Care managers see the list in their work queue, review FHIR-sourced clinical summaries, and allocate their limited time to the patients who need it most. Over six months, readmissions drop by 12 percent. The managers are not working harder; they are working smarter, guided by the intelligence the system provides.

Example 2: Revenue Cycle Optimization

A hospital's revenue cycle team struggles with claim denials. Sorting through thousands of claims manually to find those most likely to be denied is impossible. The organization deploys an AI model trained on historical claims data, clinical documentation (represented as FHIR resources), and payer patterns. The model flags high-risk claims before submission and suggests improvements to documentation.

Revenue staff review the flagged claims, verify the AI's reasoning using FHIR-linked clinical data, and make corrections or add supporting evidence. Denial rates fall by 18 percent in the first year. The AI does not replace human expertise; it amplifies it by focusing attention where it matters.

Example 3: Clinical Decision Support

A primary care practice wants to close care gaps—patients overdue for screenings, vaccinations, or chronic disease management. Manually tracking these gaps across thousands of patients is slow and error-prone. The practice implements a FHIR-based care gap engine. It queries Observation and Procedure resources to identify who is overdue, then uses an AI prioritization model to rank patients by urgency and likelihood of response.

Clinicians see the prioritized gaps at the point of care, embedded in their EHR workflow. They address them during visits or trigger outreach campaigns. Screening rates for diabetic retinopathy increased from 55 percent to 68 percent in nine months. Quality metrics improve, and the practice earns higher performance bonuses in value-based contracts.

These examples share a common structure: FHIR provides the data backbone, AI provides the intelligence, humans provide the judgment, and feedback drives continuous improvement.

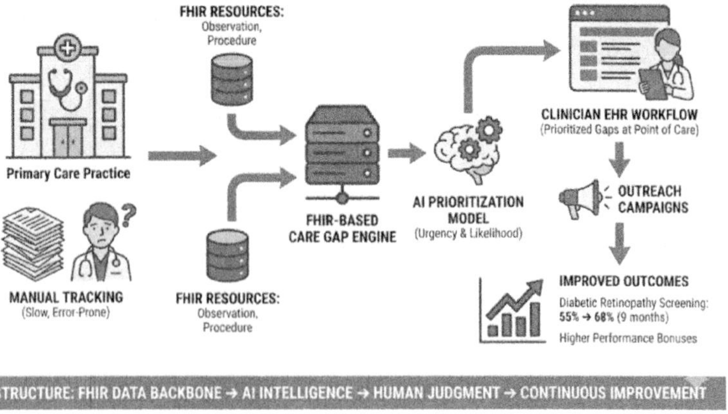

What Makes This Partnership Different from Past "Healthcare IT Revolutions"

Healthcare has seen waves of technology promises before—EHR adoption, big data analytics, population health platforms, and precision medicine. Many delivered real value, but many also fell short of expectations. Why should FHIR + AI be different?

Three things set this partnership apart:

1. FHIR is a standard, not a vendor product. Unlike proprietary integration engines or custom data warehouses, FHIR is an open, international standard. Every major EHR vendor, cloud platform, and healthcare technology company supports it. That means FHIR-based investments are more portable, more interoperable, and less likely to become technical debt when vendors change.

2. AI is increasingly accessible and embedded. A decade ago, AI required rare expertise and expensive infrastructure. Today, cloud platforms provide AI services that can be integrated via APIs. Pre-trained models can be fine-tuned for specific healthcare tasks. Tools for monitoring, explainability, and fairness are maturing. AI is moving from research labs to operational workflows.

3. Regulatory and market forces are aligned. Interoperability mandates (the 21st Century Cures Act, CMS rules, and TEFCA) are driving organizations toward FHIR adoption. Value-based contracts and quality reporting create demand for better data and intelligence.

Workforce shortages make automation and intelligent prioritization essential. The environment is forcing change in the direction that FHIR and AI naturally support.

That said, adoption is not automatic. FHIR and AI bring new complexities—data quality, model governance, privacy, fairness, security. The organizations that succeed will be those that treat FHIR + AI not as a technology project but as an operating model requiring leadership, governance, and discipline.

The Manager's Role in the Partnership

If you are reading this book, you are likely an IT leader, operations manager, clinical informatics director, or transformation lead. You are not writing FHIR profiles or training AI models yourself, but you are accountable for whether these capabilities deliver value and avoid harm.

Your role in the FHIR + AI partnership has five dimensions:

1. Connecting strategy to execution. You translate organizational goals—reduce readmissions, improve care gaps, cut administrative waste—into concrete FHIR + AI use cases. You help executives understand what is realistic, what resources are needed, and what risks must be managed.

2. Building the right foundations. You ensure your FHIR backbone is solid: mature enough to support AI, governed well enough to be trusted, and extensible enough to grow. You work with technical teams to prioritize which

FHIR resources to implement first and which data quality issues to address.

3. Governing responsibly. You establish guardrails around how AI is used: who can deploy models, what testing is required, how fairness and transparency are evaluated, and how incidents are handled. You make sure FHIR data flows respect privacy, consent, and security expectations.

4. Orchestrating across silos. FHIR and AI initiatives cut across clinical, IT, data, compliance, and operational teams. You bring these groups together, clarify roles, resolve conflicts, and keep everyone aligned with shared goals.

5. Managing change. New workflows powered by FHIR and AI disrupt established habits. You help staff understand why change is happening, train them on new tools, and create feedback channels so their concerns are heard and addressed.

None of this is easy. But the alternative—letting FHIR and AI happen in disconnected pockets without strategy, governance, or coordination—is far riskier.

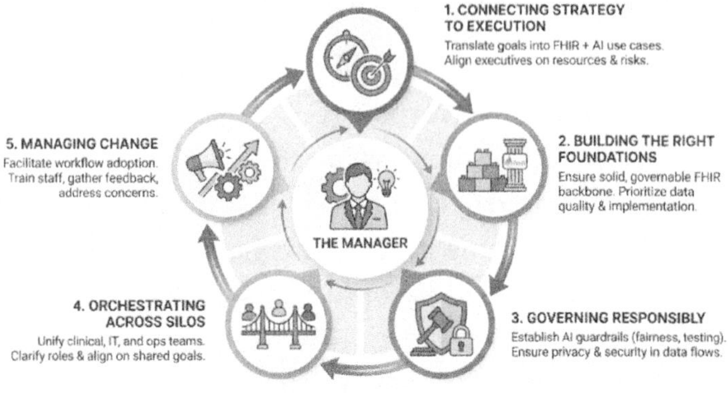

What You Will Learn in This Book

This book is your operational guide to the FHIR + AI partnership. Each chapter builds on the last, taking you from concepts to concrete action.

- **Chapter 2** establishes urgency. Why does this matter now? What are the forces—regulatory, financial, operational—pushing healthcare toward FHIR and AI?
- **Chapter 3** demystifies FHIR. What is it really? How does it work? What do managers need to know without becoming FHIR engineers?
- **Chapter 4** explains how AI turns FHIR data into action—the patterns, the use cases, the value drivers.
- **Chapter 5** shows you the symbiotic workflow in detail—how to design, implement, and refine the loop.

- **Chapter 6** focuses on governance—trust, safety, consent, and responsible use.
- **Chapter 7** covers infrastructure—how to build an AI-ready FHIR environment.
- **Chapter 8** provides a pilot-to-production pathway so you can move from experiments to enterprise capabilities.
- **Chapter 9** applies a Responsible AI lens to every stage of the lifecycle.
- **Chapter 10** addresses security—how to protect the backbone that everything else depends on.
- **Chapter 11** looks ahead to the learning health system—the long-term destination of FHIR + AI maturity.

By the end, you will have a mental model, a vocabulary, and a set of practical tools for leading FHIR + AI initiatives that deliver value, manage risk, and align with your mission.

Chapter 1 Takeaways

- **FHIR and AI are two halves of a modern healthcare engine.** FHIR standardizes data; AI turns data into intelligence. Together, they enable workflows that were previously impossible at scale.

- **This is an operational story, not just a technology one.** The partnership changes how work gets done, how decisions are made, and how organizations learn.
- **The symbiosis model—data flow, insight flow, action flow, learning flow—is the core pattern** you will use to evaluate and design FHIR + AI initiatives.
- **Managers are the bridge** between strategy and execution, between technical teams and operational reality, between innovation and governance.
- **The time to act is now.** Regulatory mandates, workforce strain, and market forces are creating both pressure and opportunity. Organizations that build FHIR + AI capabilities deliberately and responsibly will have a structural advantage.

In the next chapter, we will dig into urgency—the specific forces driving healthcare toward FHIR and AI right now, and why waiting is riskier than starting.

2 Why This Matters Now

A System Under Strain

On a Monday morning operations huddle, your dashboards all tell the same story in different ways: ED wait times creeping up, length of stay increasing on key units, denials nibbling at margins, staff satisfaction slipping another few points. Everyone around the table knows the drivers—aging populations, more complex patients, clinician shortages, regulatory obligations piling on—but no one is surprised anymore. The surprise is how normal the strain has become.

What has also become normal is the underlying chaos in the data. The ED team complains that they cannot view recent outpatient visits without logging in to another system. Care managers track high-risk patients on spreadsheets because the EHR cannot produce the list they actually need. Revenue integrity staff spend hours re-constructing documentation from clinical notes that were never designed for payers. Analytics teams scramble to glue data together for each new executive question. You end up managing the symptoms of fragmentation rather than treating the cause.

FHIR and AI enter this picture as structural tools, not magic fixes. FHIR provides a disciplined way to integrate information across systems and organizations. AI gives you a disciplined way to apply intelligence to that information at the workflow level—prioritizing, predicting, and preparing work. The combination does not eliminate the external pressures, but it changes your leverage. Instead of throwing more people at broken processes, you can redesign how work flows and how decisions get made.

The Cost of Fragmentation

Every manager in healthcare lives with fragmentation, but its costs are often invisible because they are spread out: five extra minutes per case here, a duplicated test there, a missed care gap that shows up months later. When you add those micro-costs together across thousands of patients and staff, the impact is enormous.

Fragmentation shows up when:

- A hospitalist has to click through ten screens to piece together a patient's recent history.
- A care manager cannot see claims or community data, so they over- or under-serve patients based on partial information.
- A quality team spends two months preparing a performance report that arrives too late to change behavior.
- A payer and provider argue over a denial because their records tell different stories about the same encounter.

In each case, the root cause is the same: data about the same patient, encounter, or condition lives in different places, in different formats, under different rules. The result is operational drag. People do extra work to compensate. Decisions are slower, less precise, and harder to explain.

FHIR is designed to attack this head-on. It gives you a common model—Patient, Encounter, Observation, Condition, Medication, Coverage—that can be implemented across systems. Instead of bespoke interfaces for every connection, you build a backbone once and reuse it. AI then amplifies the benefit. Once data is consistently available via FHIR, AI can scan it to identify patterns, flag anomalies, and prepare work for humans to act on quickly.

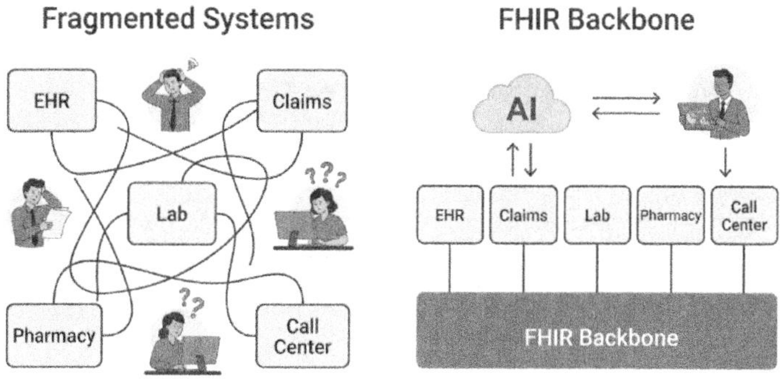

Regulatory Pressure: Interoperability Is No Longer Optional

For years, interoperability was treated as a "good to have," often deferred in favor of more immediate fires. That era is ending. Regulatory frameworks are making standardized data exchange mandatory rather than optional. Interoperability mandates are no longer about abstract policy; they show up as deadlines, audits, and potential penalties that land on your desk.

Information blocking rules expect providers and payers to make electronic health information available in usable forms, not just PDFs faxed faster. TEFCA and similar frameworks are pushing toward trusted exchange at the regional and national levels. Payer-provider rules demand that clinical and administrative data can be shared in standardized ways. Underneath all of these, FHIR is the common language that regulators and industry groups converge on.

For managers, this changes the nature of the decision. You are no longer choosing whether to invest in interoperability; you are choosing how to invest in it. You can:

- Build tactical, one-off interfaces that minimally comply and become tomorrow's technical debt.
- Or invest in a FHIR-based backbone that meets regulatory requirements and supports AI-driven workflows.

The first path keeps you in reactive mode—scrambling to meet each new rule. The second creates an asset you can reuse: once FHIR is in place, each new interoperability obligation is less painful, and each new AI use case has a ready supply of trusted data.

Workforce Reality: You Cannot Hire Your Way Out

The workforce situation in healthcare is not a temporary blip. Clinician burnout, nursing shortages, and high turnover in administrative roles are structural trends. Many organizations already struggle to fill open positions. Recruiting more people to feed existing processes is neither sustainable nor affordable.

On the clinical side, time is the most precious resource. Physicians and nurses face constant pressure to do more in less time—see more patients, complete more documentation, respond to more portal messages. On the operational side, the revenue cycle, scheduling, and call center teams are contending with increasing volumes while maintaining flat or shrinking headcount.

In this context, AI on a FHIR backbone is not a luxury; it is one of the few viable ways to create capacity without burning out the people you depend on. Consider where the time actually goes today:

- Re-entering data that already exists somewhere else.
- Manually reconciling inconsistencies between systems.
- Sorting and prioritizing tasks by hand.
- Searching across multiple tools to assemble a complete picture.

FHIR reduces the need for reentry and reconciliation by making data reusable. AI reduces some manual sorting and searching by preparing and prioritizing work. The goal is not to reduce staff to zero. The goal is to allow your existing workforce to spend a greater percentage of their time on high-value activities—clinical reasoning, relationship-building, negotiation, and problem-solving.

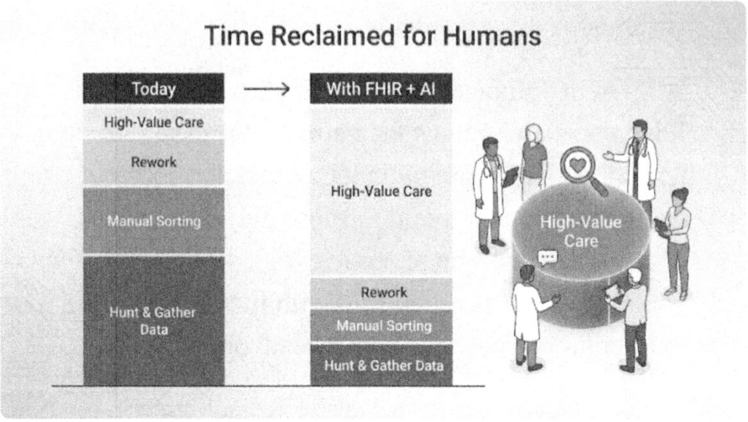

Patient and Member Expectations: Consumer Experience Comes to Healthcare

Patients and members increasingly compare their healthcare experience to the digital services they use elsewhere. They wonder why it is easier to track a package than a referral, or why they can see their banking transactions in one app but must call three offices to piece together their care plan. They are often willing to change providers or plans when these frustrations accumulate.

This shift in expectations affects you in subtle ways. Patients expect:

- Transparent access to their records through apps they choose.
- Seamless handoffs between primary care, specialists, hospitals, and community services.
- Proactive reminders and outreach, not just reactive scheduling.
- Clear explanations of why certain services are recommended or denied.

FHIR underpins many of these experiences. It enables patient-facing apps to pull data from EHRs and payers in standardized ways. It supports shared care records across settings. It allows a care journey to be stitched together from multiple systems into a coherent narrative.

AI helps personalize and optimize journeys. It can:

- Identify when a patient is likely to miss a follow-up and trigger outreach.
- Tailor messages to the patient's history and preferences.
- Highlight barriers—transportation, language, cost—that might derail a care plan.

If you do not build FHIR + AI capabilities, you will find yourself competing with organizations that can offer these experiences. In markets with choice, patients will drift toward care that feels coordinated and responsive. In public or safety-net contexts, expectations from oversight bodies and advocacy groups will align.

Financial Pressure: Margins, Risk, and Value-Based Care

Margins in healthcare are thin, and volatility is high. Small shifts in payer mix, denial rates, or value-based contract performance can make the difference between surplus and deficit. At the same time, more revenue is tied to performance metrics, risk scores, and outcomes, not just volume.

Fragmented data and manual processes make it harder to:

- Accurately capture risk and severity in documentation.
- Proactively manage populations attributed to value-based contracts.
- Reduce avoidable denials that stem from missing or inconsistent information.

FHIR and AI can change this equation. With standardized clinical and administrative data, AI models can:

- Flag documentation gaps that affect risk scores before claims go out.

- Identify which members in a value-based panel most need intervention this month.
- Predict which claims are likely to be denied and why, so staff can fix issues proactively.

This is not about gaming the system. It is about ensuring your organization is recognized and compensated for the care it actually delivers, and about using resources where they will have the greatest financial and clinical impact.

From a manager's perspective, the question is simple: do you want to continue fighting financial fires on a case-by-case basis, or invest in capabilities that systematically reduce leakage and variability? FHIR + AI is one of the few levers that can both improve care and stabilize finances.

Why "Wait and See" Is the Riskiest Strategy

Given all of this, it is tempting to say, "We'll move once the standards settle, the tools mature, and regulators clarify everything." That instinct is understandable—but dangerous.

Waiting carries three specific risks:

1. **Capability gap** – Organizations that start now will, in a few years, have mature FHIR backbones, reusable AI components, and staff experienced in these workflows. If you wait, you will be trying to catch up while they are optimizing.

2. **Compliance scramble** – Interoperability and AI-related expectations will not pause while you wait. If you postpone investing in FHIR and responsible AI, you are likely to face painful, last-minute compliance projects that deliver minimal strategic value.

3. **Cultural drift** – As vendors embed FHIR and AI into their products, pockets of usage will appear in your environment whether you plan for them or not. Without a coherent strategy, you will end up with fragmented tools and inconsistent guardrails that are much harder to rationalize later.

The alternative is not to rush mindlessly into every new technology. It is to start deliberately with a focused set of use cases, a clear FHIR roadmap, and a responsible AI posture. You do not need everything in place to begin. But you do need to begin building the muscles—technical, governance, and cultural—that will enable you to move faster and more safely over time.

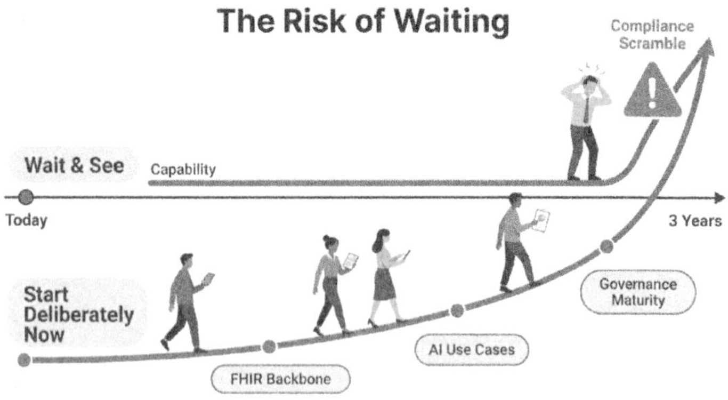

The Manager's Lens: What This Chapter Means for You

Stepping back, this chapter has laid out the forces converging on you: regulatory demands, workforce strain, rising patient expectations, and financial pressure. It has been argued that FHIR + AI is not another "nice to have" in that mix; it is one of the few levers that addresses multiple pressures at once—structurally, not superficially.

As a manager, this shapes three near-term decisions:

- **Prioritization** – You cannot do everything. FHIR + AI work should be tied to the problems that are already costing you time, money, or trust—readmissions, denials, care gaps, operational bottlenecks—not to abstract innovation goals.
- **Investment stance** – Instead of treating each interoperability or AI request as a separate project, you invest in a backbone and a set of patterns you can reuse. That means saying "yes, but on our FHIR + AI platform and within our guardrails," not "yes to every bespoke solution."
- **Narrative** – You frame FHIR and AI not as experimental technology, but as how you will meet your obligations and sustain your mission. You explain to executives, clinicians, and staff that this is about making their work more effective and sustainable, not about chasing buzzwords.

The rest of this book is about how to do that in practice. The next chapter demystifies FHIR itself, what it is, how it works, and what you need to know to oversee it without becoming a standards engineer.

3 Understanding the FHIR Framework

Why Managers Need to Understand FHIR (Without Becoming Engineers)

If AI is the brain of your modern healthcare engine, FHIR is the circulatory system. It moves standardized clinical and administrative information where it needs to go—between EHRs, payer platforms, population health tools, patient apps, and AI services. If that circulation is weak or chaotic, your AI initiatives will be fragile, and your interoperability obligations will be painful.

You do not need to memorize the FHIR specification or design implementation guides. You do need a manager-level understanding of what FHIR is, how it works structurally, and what good (and bad) FHIR usage looks like in practice. That understanding lets you ask better questions, evaluate vendor claims, prioritize investments, and connect FHIR work to real workflows instead of abstract compliance.

In this chapter, we'll translate FHIR into plain language. We'll cover the core building blocks (resources, profiles, APIs), how they show up in real workflows, and what you, as a manager, should pay attention to when someone says, "We'll do it with FHIR."

FHIR in One Page: The Essentials

FHIR stands for **Fast Healthcare Interoperability Resources**. The name carries three ideas:

- **Fast** – It is designed for web-style exchange (think modern APIs), not nightly batch files.
- **Healthcare** – It focuses on clinical and related administrative data.
- **Resources** – It breaks information into modular building blocks like Patient, Encounter, Observation, Condition, Medication, Coverage.

At a high level, FHIR does three things for you:

1. It defines *what* data looks like (resources and their fields).
2. It defines *how* to exchange that data (APIs and standard operations).

3. It provides a base for *how* to constrain and govern that data (profiles and implementation guides).

This is what makes FHIR different from previous "just another interface" projects. It is a shared language that the industry is converging on. Once you agree on that language, you can plug systems together more predictably, reuse data across workflows, and give AI a cleaner foundation to work from.

From a manager's standpoint, the key mental shift is this: you are no longer paying for one-off bridges between systems; you are investing in a backbone that supports many current and future connections.

FHIR Resources: The Lego Bricks of Healthcare Data

FHIR resources are the basic building blocks. Each resource represents a specific "thing" in healthcare with enough structure to be useful and enough flexibility to fit different contexts.

Some of the most important ones for AI-ready workflows are:

- **Patient** – Who the person is. Demographics, identifiers, contact info.
- **Encounter** – When and where the person received care. Visits, admissions, discharges.

- **Condition** – What the person is being treated for. Diagnoses and problems.
- **Observation** – Measurable facts. Vitals, lab results, scores, assessments.
- **MedicationRequest / MedicationStatement** – What medications are ordered or taken.
- **Procedure** – What was done. Surgeries, imaging, interventions.
- **Coverage** – How the care is paid for. Payers, plans, member IDs.

Each resource has defined fields and relationships. For example, an Encounter links to a Patient; Observations can link to Encounters; Conditions can have onset dates and statuses. AI models and analytics rely on this structured representation to understand context: not just "what lab value," but "for which patient, in which visit, with what diagnoses and medications at the time."

As a manager, you don't need to know every field, but you should know which resources matter most for your priority use cases. For readmissions, that might be Patient, Encounter, Condition, Observation, MedicationRequest, Coverage. For denials, you might add Claim and Explanation of Benefit. Knowing these anchors helps you have grounded conversations about scope and readiness.

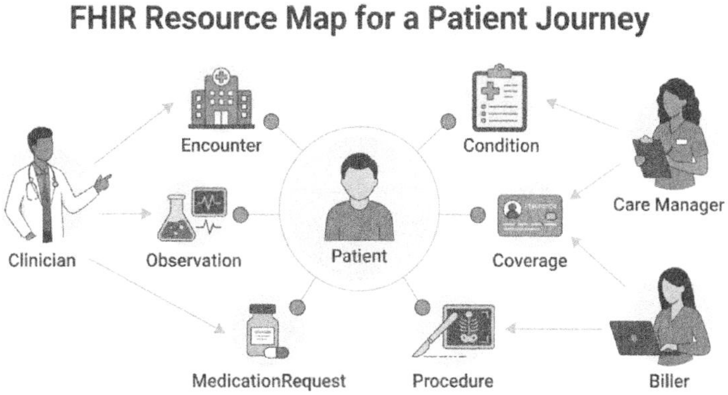

Profiles and Implementation Guides: From Generic to "How We Do It Here"

Out of the box, FHIR resources are deliberately flexible. That flexibility is powerful, but if every organization uses different subsets of fields, coding systems, and rules, interoperability and AI both suffer. You need a way to say, "For our blood pressure measurements, we always send these fields, using this code system, in this way."

That's where **profiles** and **implementation guides (IGs)** come in:

- A **profile** takes a base resource (like Observation) and tightens it—making some fields required, fixing coding systems (e.g., LOINC for certain labs), and sometimes adding extensions.
- An **implementation guide** packages a set of profiles, rules, and examples for a particular use case or

program (e.g., US Core, a quality program, a payer integration).

Think of base FHIR as a toolbox, profiles as your customized toolsets for specific jobs, and implementation guides as the instruction manuals for teams.

From a manager's perspective, profiles and IGs matter because they:

- Determine **data consistency** across sites and vendors.
- Affect **AI quality**, since models trained on inconsistent data are brittle.
- Carry **governance decisions**—for example, which fields are allowed to carry certain kinds of sensitive information.

You should know which profiles your organization relies on (e.g., US Core, specific vendor IGs, custom internal profiles) and who owns decisions to change or extend them. Uncontrolled "profile sprawl" can quietly undermine your backbone.

FHIR APIs: How Systems and AI Actually Talk

FHIR is not just data definitions; it's also about how systems exchange that data in real time. That happens through APIs—web endpoints that allow authorized clients to read and write FHIR resources.

Common patterns include:

- **Read** – "Give me this patient's Observations for the last 30 days."
- **Search** – "Give me all Encounters for this patient in the last year."
- **Create/Update** – "Record this new Condition" or "Update this Encounter."
- **Subscriptions/Notifications** – "Tell me when a new Encounter is created for this patient group."

AI systems often sit on the consuming side of these APIs. For example:

- A risk model calls FHIR APIs nightly to pull updated Encounter and Observation data to refresh scores.
- A care gap engine subscribes to changes in relevant labs or procedures and updates its gap list.
- A documentation assistant reads recent FHIR resources to draft notes, then writes AI-generated summaries back as Documents or Observations.

For you, the questions are less about protocol details and more about **how** and **how much**:

- Which systems are allowed to call your FHIR APIs, and for what purposes?
- Are API usages logged, monitored, and limited to what is necessary?

- Do AI vendors use your FHIR APIs in a way that fits your architecture, or are they pulling data out into opaque side systems?

Understanding the basic API model lets you see where risk, cost, and opportunity actually sit.

How FHIR Shows Up in Real Workflows

FHIR can sound abstract until you see where it touches everyday work. Here are a few common patterns.

Patient Access Apps
A patient uses a mobile app to see their records from multiple providers. Behind the scenes, the app calls each provider's FHIR APIs, pulling Patient, Encounter, Observation, and Medication resources into a unified view. For you, this may be a regulatory requirement, but it also serves as a foundation you can reuse later for internal AI-driven engagement.

Care Management Platforms
Your care management tool connects to an EHR via FHIR, pulling relevant Patient, Encounter, Condition, and Observation data to populate care manager queues. AI runs on that integrated dataset to prioritize outreach. Every time a care manager documents an outcome, the information can be written back in a structured format for reuse.

Quality Reporting and Value-Based Programs
Rather than manual chart abstraction, quality measures can be calculated from FHIR data feeds. Observations,

Procedures, Conditions, and Encounters flow into a measure engine. AI helps identify outliers or documentation gaps. You meet reporting obligations while improving the data you need for operational AI.

Prior Authorization and Utilization Management

FHIR can carry structured clinical evidence to payers. AI helps assemble the right data and predict likely determinations. Over time, patterns in FHIR-based submissions and payer responses feed back into both your documentation practices and your AI models.

In all of these, FHIR is not a separate "project"; it is the plumbing underneath tools people actually use. When done well, staff may not even realize FHIR is involved—they notice that information is where it should be.

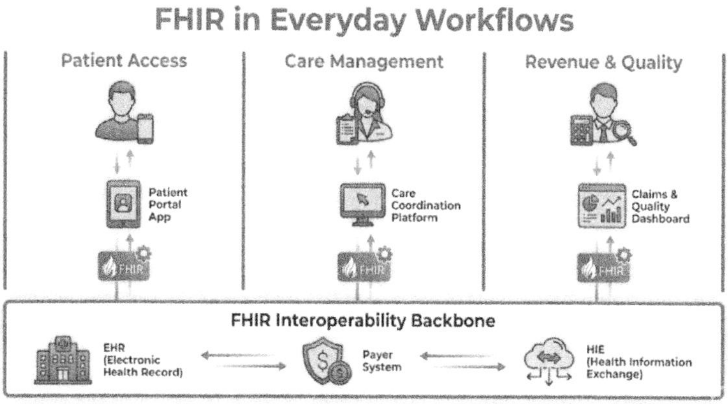

FHIR and AI: Why Structure Matters So Much

AI thrives on structured, consistent data. FHIR is how you deliver that. Without FHIR (or something like it), AI teams often spend most of their time wrangling data—extracting data from EHRs, normalizing fields, mapping codes, and handling missing values. Each new AI use case becomes a new plumbing project.

With a mature FHIR backbone, AI work changes character:

- Feature engineering can reuse established FHIR fields and codes.
- Models can be applied across sites because the data is consistently structured.
- Monitoring and governance can attach to a known schema instead of a tangle of custom tables.

Consistency matters especially for:

- **Fairness and bias checks** – You can compare outcomes across demographic fields if those fields are reliably captured in Patient and related resources.
- **Explainability** – You can show clinicians which Observations or Conditions influenced a prediction because those inputs are standardized and linkable.
- **Drift monitoring** – You can detect changes in data distributions (e.g., lab values, procedure mix) over time when they are represented consistently.

The more your AI portfolio grows, the more you will feel the difference between "data plus models" versus

"FHIR-structured data plus models." The second is simply more scalable and governable.

What "Good FHIR" Looks Like (and Red Flags for Managers)

You don't have to review FHIR payloads yourself, but you should know what good practice looks like and what warning signs to watch for.

Good FHIR characteristics:

- **Coverage of key resources** for your use cases is clearly documented (e.g., "We support US Core Patient, Encounter, Observation, Condition, Medication, Coverage for inpatient workflows").
- **Profiles are defined and owned**, with clear stewards who can explain what fields are required and why.
- **Code systems and value sets are standardized** (e.g., LOINC for labs, SNOMED/ICD for diagnoses), not ad hoc free text everywhere.
- **APIs are secured and monitored**, with access aligned to least privilege and business purpose.
- **FHIR is reused** across multiple projects (AI, reporting, external exchange), not recreated differently for each one.

Red flags:

- Vendors or internal teams saying "we support FHIR" but are unable to specify which versions, resources, or profiles.
- Heavy reliance on "extensions" for basic concepts that should exist in core resources, suggesting either poor design or misalignment with standards.
- Multiple teams building parallel FHIR layers with different conventions, creating "interoperability silos" inside your own organization.
- No clear governance around who can change profiles or approve new data uses, leading to quiet drift in meaning and quality.

When you see red flags, you don't need to fix them yourself. You do need to ask for a plan: who will rationalize profiles, align with relevant guides, and ensure FHIR is an asset rather than a patchwork?

Your Role: Steering, Not Specifying

In the end, your job is not to design FHIR schemas or write API endpoints. Your job is to ensure that FHIR is:

- **Strategically aligned** – implemented in ways that support your highest-value use cases and regulatory obligations.
- **Governed and owned** – with clear accountability for quality, security, and change management.

- **Connected to AI work** – so that every serious AI initiative either uses the backbone or pushes it forward, rather than bypassing it.

Practically, that means you:

- Ask architects and informatics leaders to map FHIR capabilities to your top AI and interoperability priorities.
- Insist that major AI or analytics initiatives explain how they will use FHIR data and, where necessary, help strengthen it.
- Support governance structures that make FHIR decisions transparent and repeatable, not ad hoc.

You don't need to win arguments about individual fields. You need to set expectations that FHIR is your **data operating model** for healthcare, and that AI must work with that model, not around it.

Chapter 3 Takeaways

- FHIR is more than an interface spec; it is the backbone for how your organization represents and exchanges healthcare data.
- Resources, profiles, and APIs are the building blocks that shape both interoperability and AI readiness.
- Good FHIR practice—consistent resources, clear profiles, secure APIs—directly affects the quality, fairness, and explainability of AI systems.

- As a manager, your role is to ensure FHIR is strategically used, well-governed, and tightly coupled to the AI work that matters most.

In the next chapter, we'll zoom in on AI itself: how it turns FHIR data into predictions, prioritizations, and workflow support that your teams can actually use on Monday morning.

4 How AI Turns FHIR Data into Action

From Raw Data to Useful Decisions

In the last chapter, you saw FHIR as your data backbone—the way you represent patients, encounters, labs, medications, and more in a consistent, reusable way. On its own, that backbone solves some problems: fewer re-entries, cleaner integrations, faster reporting. But the real operational leverage shows up when you layer **AI** on top of FHIR.

AI is how you move from "we have the data" to "we know what to do next." It turns standardized FHIR streams into predictions, prioritizations, classifications, and drafts that fit directly into workflows. A readmission risk score is not just a number; it is a prioritized list of patients for your care team to address. A denial risk prediction is not just analytics; it is a focused worklist for your revenue staff. A note drafted from FHIR data is not just text; it is time given back to a clinician.

In this chapter, we'll stay firmly at the manager's level. You will not learn neural network architectures. You will learn the **four core AI patterns** that matter operationally, how they depend on FHIR, and how to recognize when you're using them well—or badly—in real workflows.

The Four AI Patterns You Actually Use

Most healthcare AI use cases boil down to a small set of reusable patterns:

- **Prediction** – Estimating the likelihood of a future event (readmission, deterioration, denial, no-show).
- **Prioritization** – Sorting a list so that the most important items surface first (panels, queues, worklists).
- **Classification** – Assigning labels or categories (risk tiers, care gaps, document types, routing destinations).
- **Generation** – Producing new content based on existing data (draft notes, patient letters, summaries, messages).

These patterns often combine, but you can usually point to one as the main purpose of a given tool.

For each pattern, FHIR provides structured inputs such as Patient, Encounter, Condition, Observation, MedicationRequest, and so on. AI transforms those inputs into something that changes how people work. Your job is to ensure that the transformation is:

- **Mission-aligned** – helping with problems that actually matter.
- **Workflow-aware** – delivered at the right time and place in the process.

- **Responsibly governed** – transparent enough, fair enough, and monitored enough for the risk involved.

You don't need different governance for every model; you need a clear view of which pattern you're using and what's at stake.

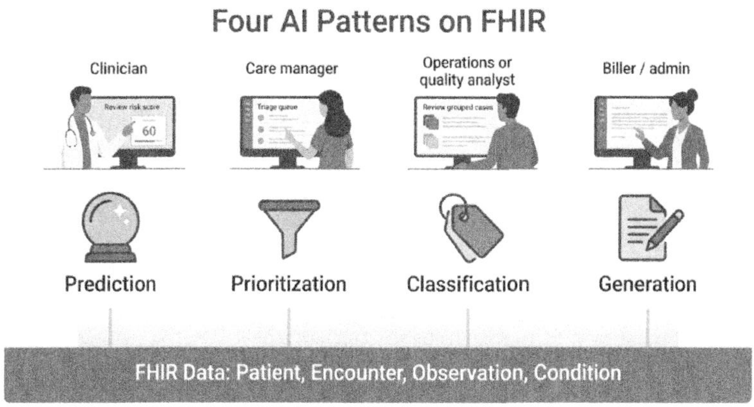

Prediction: Seeing Problems Before They Hit

Prediction models take FHIR-structured data and estimate the probability of future events. In healthcare, that often means:

- The chance a patient will be readmitted within 30 days.
- The risk of deterioration on a ward.
- The likelihood that a member will have a high-cost event in the next year.
- The probability that a claim will be denied.

From a FHIR perspective, prediction models usually consume:

- **Patient** (demographics, key risk factors).
- **Encounter** (visit history, acuity, length of stay).
- **Condition** (diagnoses, comorbidities).
- **Observation** (vitals, labs, scores).
- **MedicationRequest / MedicationStatement** (treatment patterns).
- Sometimes **Coverage** and other context.

The output is a **risk score** or probability. On its own, that score is not useful. It becomes useful when you:

- Connect it to a specific **decision** ("Who gets a care manager call?" "Which claims get extra review?").
- Present it in a **context** people trust (embedded in the EHR, a care management tool, or a revenue dashboard).
- Pair it with **next steps** ("Review these patients this morning," "Check documentation before submitting").

Key questions for you when dealing with prediction:

- What precise decision is this prediction supposed to support?
- What FHIR resources is it using as input, and are they reliable enough?

- How do we monitor whether predictions are accurate and fair over time?

Prioritization: Turning Lists into Actionable Queues

If prediction is about estimating "how likely," prioritization is about deciding "in what order." Healthcare is full of long lists:

- Panels of patients with chronic conditions.
- Queues of pending referrals.
- Inboxes of documentation or claims.
- Backlogs of tasks in care management or revenue cycle.

Without help, staff often work those lists in arbitrary ways—first-in, first-out, alphabetically, or based on who shouts loudest. AI on FHIR can change that by **scoring and sorting** items based on urgency, risk, or impact.

Examples:

- A care management queue is ordered so that patients with the highest risk and most modifiable factors appear first.
- A list of claims sorted by likelihood of denial *and* financial impact.

- A backlog of prior authorizations ordered by clinical risk and time sensitivity.

FHIR provides the data that informs those scores: Encounter history, Observation trends, Condition complexity, Coverage details, and more. The AI layer combines that into a prioritization signal. The workflow layer shows staff a **ranked queue** instead of a raw dump.

Your role is to ensure that:

- The prioritization logic is **transparent enough** that staff can understand the basics ("why is this case at the top?").
- The queue design fits **how people actually work** (batch work vs. ad hoc pick-lists).
- You have **guardrails** to avoid systematically deprioritizing certain populations or low-voice groups.

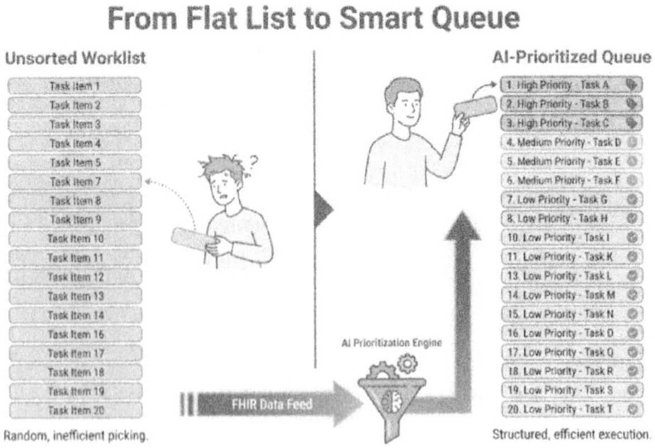

Classification: Labeling and Routing at Scale

Classification models take FHIR-linked data and assign labels or categories to it. In healthcare, this shows up as:

- **Risk tiers** (low, medium, high) for patients or episodes.
- **Care gaps** (present/absent) based on guidelines.
- **Document types** ("discharge summary," "consult note," "operative report") for routing and indexing.
- **Routing decisions** ("send this referral to this clinic," "assign this case to this team").

Classification can be rule-based ("if X and Y, then high-risk") or learned from data. Either way, it relies heavily on FHIR:

- Conditions and Observations to assess clinical status.
- Encounter and Procedure history to determine what has been done.
- Demographic and coverage details to apply the right rules.

The output is usually a **label** or a **flag**. Its value comes from:

- How reliably it reflects reality (is "high-risk" meaningful?).

- How well it feeds into **downstream workflows** (does "care gap present" trigger outreach?).
- How clearly it is documented and visible (can auditors and clinicians see *why* that label was applied?).

For you, key considerations include:

- Who owns the definition of each label? (e.g., what exactly "high-risk" means).
- How do labels appear in systems—are they clutter, or do they genuinely guide action?
- How are misclassifications found and corrected, and does that feedback improve the underlying logic?

Generation: Drafting the Work, Not Doing the Job

Generative AI—especially large language models—has opened a new pattern: **generation**. These systems can create text, and sometimes images or code, based on prompts and context. In a FHIR environment, that context often comes from standardized patient data.

Common use cases:

- Drafting clinical notes from recent Encounters, Conditions, Observations, and MedicationRequests.
- Generating plain-language patient letters explaining care plans or test results.

- Summarizing long histories into "one-page" views for clinicians.
- Drafting prior authorization narratives using FHIR-linked clinical evidence.

The key here is that the generation should **draft the work**, not make final decisions. The human remains responsible for:

- Reviewing content for accuracy and tone.
- Correcting mistakes or hallucinations.
- Deciding what gets sent, signed, or stored.

FHIR plays two critical roles:

- It **grounds** the model in real, structured facts, reducing hallucinations.
- It provides a way to **store and trace** generated content, keeping the linkage between the draft and source data.

Your questions as a manager:

- Where does generative AI save time without compromising safety? (Documentation, letters, summaries).
- How do we clearly define **human-in-the-loop** responsibilities?
- How do we prevent sensitive information from leaking via prompts or outputs?

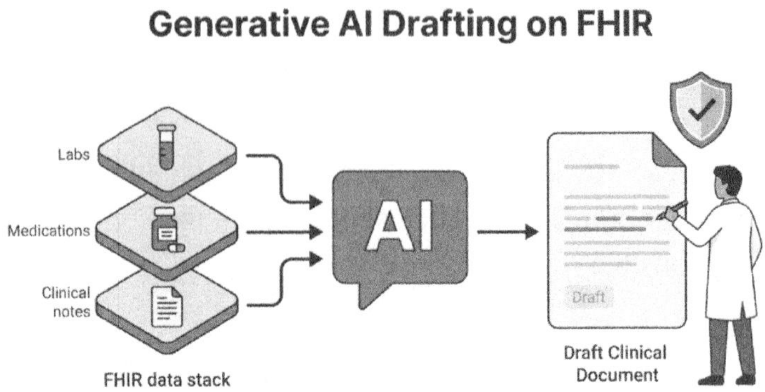

How These Patterns Combine in Real Use Cases

In practice, a single initiative often uses multiple patterns together. Consider a **readmission reduction program**:

- **Prediction** estimates each patient's risk of readmission using FHIR data.
- **Classification** assigns risk tiers (low/medium/high) and care gap flags.
- **Prioritization** sorts the care management queue by risk and modifiability.
- **Generation** drafts outreach scripts or follow-up summaries based on patient history.

Or a **denial prevention effort**:

- **Prediction** estimates the likelihood of denial for each claim.
- **Classification** labels claims by denial reason categories.
- **Prioritization** orders worklists by denial risk x dollar value.
- **Generation** drafts appeal letters and documentation summaries.

The FHIR backbone ensures that all these steps share a consistent view of the patient, encounter, and documentation. The AI patterns, stitched together wisely, change how a team spends its day.

When evaluating proposals, you can ask:

- Which patterns are we using here, and do we really need all of them?
- Where exactly does FHIR come in—what resources are we reading and writing?
- Where, in the workflow, will people see and respond to these outputs?

Where AI on FHIR Goes Wrong (and How to Spot It Early)

AI on FHIR can fail in predictable ways. Early recognition lets you correct course before damage is done.

Common failure modes:

- **Data mismatch** – Models assume fields are populated and coded consistently, but your FHIR implementation is patchy. Predictions and classifications become noisy or biased.
- **Workflow misfit** – Outputs appear at the wrong time or place (e.g., alerts firing during the wrong part of a visit), so staff ignore them.
- **Over-automation** – Human oversight is reduced to a checkbox; people feel pressured to accept AI suggestions they don't trust.
- **Opaque logic** – Staff cannot see why a patient was flagged or why a claim was prioritized, so they disengage or push back.
- **Unmonitored drift** – Models built on last year's patterns quietly degrade as populations, practices, or coding change.

Red flags you can look for:

- "We'll just plug the model in and see what happens" with no clear plan for monitoring or rollback.
- Lack of clarity about which FHIR resources and profiles a model depends on.
- No documented owner for a model in production.
- Feedback channels from users that go nowhere ("we've been complaining about this for months").

When you see these, your task is not to debug models. It is to insist on a **minimal structure**: clear inputs, clear ownership, clear metrics, clear escalation paths.

The Manager's Monday-Morning Playbook for AI on FHIR

To make this concrete, here is a playbook you can use whenever an AI-on-FHIR idea surfaces:

- **Clarify the pattern.** Ask: Is this primarily prediction, prioritization, classification, or generation? That tells you what kind of oversight you need.
- **Anchor in FHIR.** Ask: Which FHIR resources and profiles will this use? Are they mature and governed? If not, what do we need to fix first?
- **Define the decision.** Ask: What decision or action will this output change, and for whom? If that's fuzzy, the project is not ready.
- **Locate the workflow touchpoint.** Ask: Where, in which tool, and at what moment, will a human see and respond to this? Screenshots and mockups help here.
- **Set basic guardrails.** Ask: How will we monitor performance and fairness? Who owns this model in production? What's our rollback plan if it misbehaves?

- **Plan for feedback.** Ask: How can users report issues, and how will those reports change the system? If the answer is "email someone," you probably need a better mechanism.

You can apply this playbook even if you never look at a single line of code. It turns AI from a buzzword into a manageable set of patterns sitting on a data backbone you understand.

Chapter 4 Takeaways

- AI on FHIR is best understood through four patterns: prediction, prioritization, classification, and generation.
- FHIR provides the structured inputs and outputs that make these patterns reliable, explainable, and reusable.
- The value of AI comes from changing **decisions and workflows**, not just producing scores or text.
- As a manager, your job is to identify which pattern you're using, ensure FHIR readiness, tie outputs to real decisions, and demand basic guardrails and feedback loops.

In the next chapter, we'll bring everything together in the **Symbiosis Model**—a practical workflow-level view of how FHIR and AI interact with humans in a continuous loop of data, decisions, and learning.

5 The Symbiotic Workflow — Humans, FHIR, and AI in the Loop

Why You Need a Workflow, Not a Wish List

You now have the pieces: FHIR as the backbone, AI as the intelligence layer, and a clear sense of why this matters. But without a **workflow view**, those pieces stay abstract. Teams end up with scattered pilots, clever models no one uses, and FHIR feeds that power a few reports but not day-to-day work.

Managers don't manage models or data formats; they manage workflows. You care about how a discharge actually happens, how a claim actually gets paid, how a care gap actually gets closed. The symbiotic workflow is your map: it shows how **data**, **AI**, and **humans** interact in real time, in an ordered way.

This chapter puts that map in your hands. You'll see the core loop—data flow, insight flow, action flow, learning flow—and how to apply it to real use cases. You'll also see where this loop breaks and how to design it so humans stay in command, not just "in the loop" on paper.

The Symbiosis Loop: Four Stages

The symbiotic workflow can be described in four stages:

1. **Data Flow (FHIR)** – Reality is captured as standardized data.
2. **Insight Flow (AI)** – Data is turned into predictions, priorities, labels, and drafts.
3. **Action Flow (Humans)** – People decide and act based on AI-augmented information.
4. **Learning Flow (Feedback)** – Outcomes feed back to improve both models and processes.

If you remember only one diagram from this book, make it this loop. Every serious FHIR + AI initiative should be able to explain how it moves through these four stages, and where your people fit in.

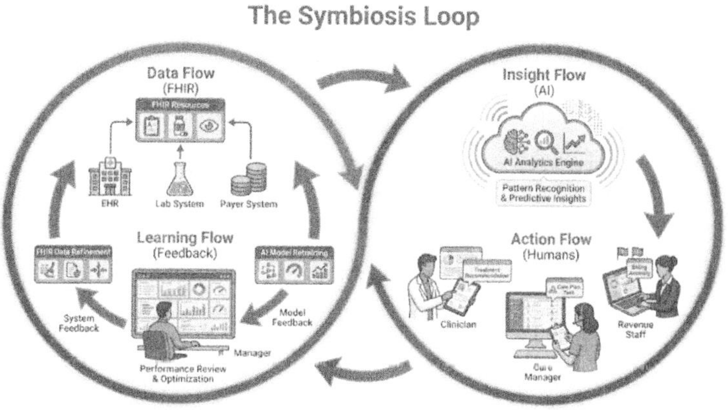

Stage 1: Data Flow — Getting Reality into FHIR

Everything starts with what actually happens to patients and members—appointments, admissions, vitals, labs, prescriptions, procedures, payments, and outreach. The **data**

flow stage is about capturing that reality into FHIR in a way that is:

- Timely enough for decisions (hours or minutes, not months).
- Complete enough for AI to see meaningful patterns.
- Structured enough to be reused consistently.

In practical terms, this includes:

- EHR events being exposed as FHIR resources (Patient, Encounter, Condition, Observation, MedicationRequest, Procedure).
- Payer and administrative systems mapping coverage and claims-related data into FHIR where needed.
- External sources—HIEs, registries, some devices—integrating via FHIR when feasible.

From a manager's lens, key questions at this stage:

- For this workflow, which FHIR resources matter most, and are they reliable?
- How quickly does new data appear in FHIR—near real time or long after the fact?
- Who owns the quality and completeness of those resources?

If the data flow is weak—missing fields, inconsistent coding, delayed updates—the rest of the loop will wobble, no matter how good your AI team is.

Stage 2: Insight Flow — FHIR into AI

Once data is flowing, the **insight-flow stage uses AI to transform FHIR-structured information into insights that change** how people see their work. In Chapter 4, we described the four main patterns: prediction, prioritization, classification, and generation.

At this stage, your teams are:

- Selecting which FHIR resources and fields feed each model.
- Engineering features (e.g., "number of admissions in 6 months") from those resources.
- Training and validating models, or configuring rules, to produce useful signals.
- Deciding how to represent those outputs (scores, tiers, flags, drafts) in systems.

The critical management move here is to insist that **insights are anchored back in FHIR and workflows**, not trapped in a black-box model server. That means:

- Writing AI outputs back as FHIR Observations, Flags, or other appropriate resources when feasible.
- Carrying metadata about how the insight was produced (model version, timestamp, key contributing factors).
- Designing simple, human-readable summaries of "why" along with the raw score or label.

This is where you make explainability and traceability operational. If you ever need to answer "why was this patient flagged?" or "why was this claim prioritized?", your insight flow design determines whether you can provide that insight.

Stage 3: Action Flow — Where Humans Actually Use It

The **action flow** stage is where most FHIR + AI efforts succeed or fail. You can have beautiful data and impressive models, but if outputs land in the wrong place, at the wrong time, in the wrong format, people will ignore them.

You need to decide, for each use case:

- **Who** is expected to act (clinician, care manager, scheduler, revenue analyst, supervisor).
- **Where** they will see the AI-augmented information (EHR screen, care management tool, inbox, huddle report, queue).
- **When** in their workflow, they should see it (before a visit, during documentation, at daily huddle, at claim submission).
- **How** they can respond (accept, override, snooze, escalate, annotate).

Examples:

- A care manager sees a prioritized list of patients every morning, with risk scores and key drivers shown inline.
- A hospitalist sees a risk banner in the EHR on admission, with drill-down into recent FHIR Observations and Conditions.

- A biller sees denial risk flags and suggested fixes before submitting a claim.

Your job is to ensure this stage is **co-designed** with the people who will live in it. That means real screens, real workflow walk-throughs, and realistic scenarios—not just a technical spec that says “display risk score.

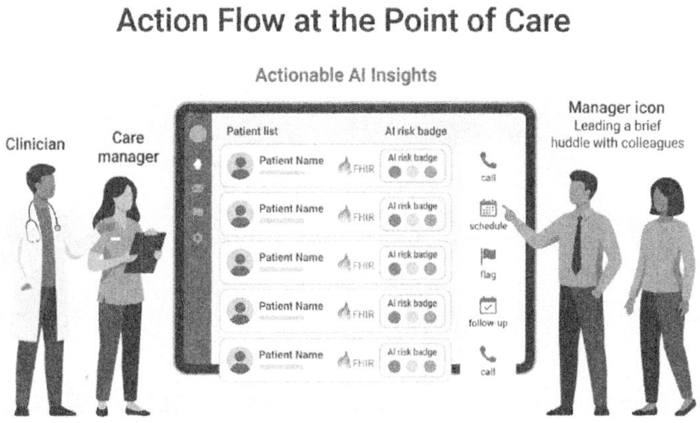

”

Stage 4: Learning Flow — Closing the Loop

Without a **learning flow**, your symbiosis loop becomes a one-way conveyor belt: data in, AI out, humans act, and then… nothing. You miss the chance to see whether the partnership is actually working and to improve it over time.

The learning flow captures:

- **What happened after the action** – Did the patient improve? Did the claim get paid? Did the care gap close?

- **How humans responded** – Did they agree with the AI? Override it? Ignore it? Why?
- **How the environment changed** – New guidelines, new patient demographics, new payer behaviors.

Technically, this means:

- Outcomes and actions are documented in systems that feed back into FHIR (e.g., updated Conditions, Encounters, Observations, Claim outcomes).
- Overrides and disagreements are logged in structured ways, not lost in free text.
- Monitoring views track performance over time: accuracy, fairness, impact on outcomes, and workload.

From your seat, the learning flow shows up as:

- Regular reviews of key workflows with simple dashboards.
- Adjustments to models, thresholds, and workflows based on evidence.
- Decisions to scale, fix, or retire AI-augmented workflows.

This is where you move from "we launched something" to "we operate this capability and continuously refine it."

5.7 Applying the Symbiosis Loop to a Concrete Use Case

Let's walk the loop for a specific example: **reducing 30-day readmissions on a medicine unit**.

Data Flow (FHIR)

- Patient, Encounter, Condition, Observation, MedicationRequest, and Coverage data are captured in FHIR from the EHR and relevant systems.
- The FHIR layer reliably reflects admissions, comorbidities, vitals, labs, meds, and basic social factors within hours.

Insight Flow (AI)

- A model uses the FHIR data to predict readmission risk at several points: admission, mid-stay, and pre-discharge.
- It outputs a risk score and risk tier, plus key drivers (e.g., prior admissions, lab trends, specific diagnoses).
- The risk tier is written back as an FHIR Observation or Flag linked to the relevant Encounter.

Action Flow (Humans)

- Each morning, the care team huddle uses a list of inpatients with their readmission risk tiers visible.
- High-risk patients are discussed: do we need extra education, closer follow-up, social work consults?
- Care managers receive a prioritized queue of high-risk discharges to arrange post-discharge calls and appointments.

Learning Flow (Feedback)

- Readmission outcomes are tracked via new Encounters and associated Conditions.

- Care managers document whether outreach was completed and what barriers they encountered.
- Periodic reviews compare risk predictions to actual readmissions, by unit and patient subgroup, and adjust models, thresholds, or workflows accordingly.

You can map other use cases—care gap closure, denial reduction, ED throughput—using the same four-stage template. If any stage is missing or vague, you have a design problem to fix before scaling.

Where the Symbiosis Loop Breaks

Most FHIR + AI disappointments can be traced to **break points** in this loop:

- **Data Flow break** – Key data never makes it into FHIR, or arrives too late. The model is starved or biased from the start.
- **Insight Flow break** – AI is built on a sidecar dataset unrelated to your FHIR backbone, making it hard to explain or maintain.
- **Action Flow break** – Outputs are dumped into a dashboard nobody uses, or alerts fire at irrelevant moments, leading to alert fatigue.
- **Learning Flow break** – No one monitors results, so drift, bias, or simple ineffectiveness go unnoticed.

Symptoms you might see:

- "The risk score isn't trusted" (often an Action or Data break).
- "We don't really know if this is working" (a Learning break).
- "Every new AI project seems to start from scratch" (a Data/Insight break with no reusable patterns).

When you see these, use the loop as a diagnostic tool. Ask, stage by stage: what exactly happens here? Who owns it? What evidence do we have that this stage is working?

Designing with Humans at the Center

A common trap is treating "human in the loop" as a box you check by putting a person in after the model. In practice, **human-AI collaboration** is much more nuanced. You need to consider:

- **Control** – What decisions remain fully human? Which are AI-assisted? Which, if any, are automated with human override?
- **Cognitive load** – Does the AI reduce or increase the mental effort required? Are explanations quick and meaningful, or do they add reading burden?
- **Authority** – Do staff feel they can disagree with the AI without penalty? Are overrides truly possible in the time they have?

Designing collaboration through the symbiosis loop helps. For each stage, you can ask:

- Data Flow – Are we capturing what humans actually notice and do, not just what the system already logs?
- Insight Flow – Are we turning data into insights that match how humans think about risk, priority, and categories?
- Action Flow – Do we present insights at the moment of decision, and do we make the options clear?
- Learning Flow – Are we capturing human disagreement, frustration, and workarounds as signals to improve the system?

If your loop respects human judgment and workflow realities, adoption follows. If it doesn't, people will quietly route around it.

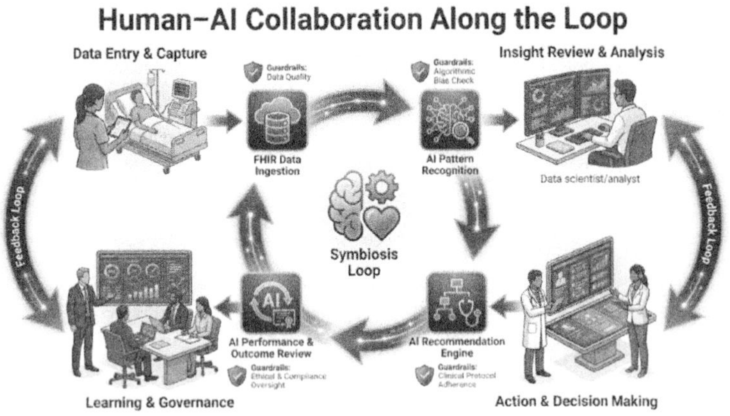

The Manager's Symbiosis Checklist

You can use this checklist whenever a FHIR + AI initiative is proposed—or when you're trying to rescue one that feels stuck:

Data Flow

- Which FHIR resources power this workflow?
- Are they complete and timely enough? If not, who owns fixing that?

Insight Flow

- Which AI pattern dominates (prediction, prioritization, classification, generation)?
- Are insights written back into a form we can trace and monitor?

Action Flow

- Who sees the outputs, where, and when?
- How do they respond, and what are their options (accept, override, escalate)?

Learning Flow

- How will we track outcomes and overrides?
- Who reviews this information, how often, and what kinds of changes can they make?

If you can't answer these questions clearly, the workflow is not ready for prime time. If you can, you've turned a vague "we'll use AI on FHIR" idea into a concrete, governable loop.

Chapter 5 Takeaways

- The symbiotic workflow—data, insight, action, learning—is the core operating model for FHIR + AI.

- Each stage has clear owners, failure modes, and design choices that fall squarely in a manager's domain.
- Human-centered design along this loop determines whether FHIR + AI becomes background infrastructure that quietly powers better care—or just another set of unused tools.

In the next chapter, we'll layer governance directly onto this loop, showing how trust, safety, consent, and responsible AI practices are built **into** the workflow rather than bolted on afterward.

6 Governance by Design — Trust, Safety, and Consent

When "Just Ship It" Stops Working

A regional health system proudly rolls out its first AI-assisted discharge planning tool. It pulls structured data via FHIR, predicts readmission risk, and nudges care teams to arrange follow-up for high-risk patients. For the first few weeks, everything looks great: staff like the prioritized lists, readmission rates dip in pilot units, and leadership celebrates a modernization win.

Then a pattern emerges. A physician raises concerns that patients with limited English proficiency seem to be flagged less often than others, even when they have similar clinical risks. A care manager reports that one unit is overriding almost all of the AI's recommendations because they "don't trust the score." A privacy officer asks whether patient consent covers the use of this data for AI beyond direct care. No one can clearly explain how the model was trained, who approved its use, or what the escalation path is when something seems off.

The problem is not that someone used FHIR or AI. The problem is that they did so **without governance-by-design**. The system works technically, but no one is sure it works safely, fairly, or under the right consent and oversight. You end up in reactive mode—trying to fix governance after deployment, under pressure.

This chapter is about avoiding that trap. Governance by design means building trust, safety, and consent into your FHIR + AI workflows from the beginning, so you can say "yes" to more ideas without losing control.

Governance as Enabler, Not Brake

Governance has a reputation problem. Many teams hear the word and think "slow," "bureaucratic," or "no." In an era of AI hype and tight deadlines, it can feel tempting to treat governance as optional overhead. But in FHIR + AI work, the opposite is true: **without governance, you eventually go slower**.

Ungoverned AI on loosely governed data creates:

- Rework when a model must be pulled back after an incident.
- Distrust among clinicians and staff who never felt consulted.
- Regulatory exposure when auditors ask questions you cannot answer.
- Fragmented one-off rules that differ across teams and vendors.

Governance by design does something different. It gives you **operational clarity**:

- Clear criteria for which FHIR + AI ideas move forward.

- Pre-agreed safeguards for higher-risk workflows.
- Standard ways to document data use, model behavior, and human oversight.
- Known escalation paths when something goes wrong.

With that clarity, you can approve low-risk experiments quickly, concentrate heavier oversight where it's actually needed, and show leadership and regulators how you manage AI and FHIR responsibly.

Governance is not a separate track from your symbiosis loop. It sits on top of it and inside it, shaping each stage.

The Governance Layer on the Symbiosis Loop

Think back to the four stages: **data flow, insight flow, action flow, and learning flow**. Governance lives in each:

- On **data flow**, governance defines which data can flow where, under what consent and access rules.
- On **insight flow**, it sets expectations for fairness, transparency, and acceptable error levels.
- On **action flow**, it clarifies who is accountable for decisions and how overrides are handled.
- On **learning flow**, it requires monitoring, incident response, and continuous improvement.

You can picture governance as a set of rings wrapped around the loop, with specific questions at each stage.

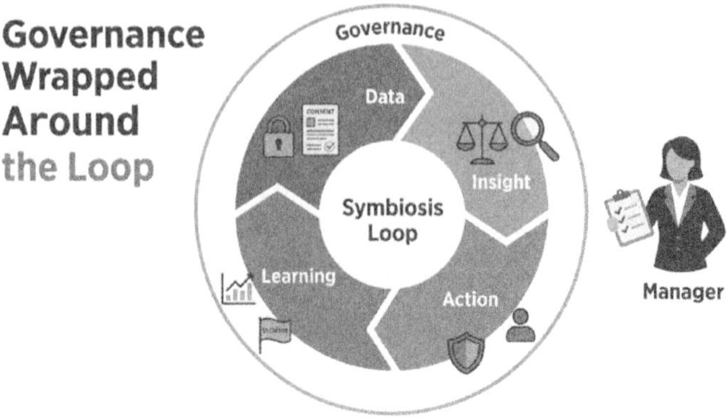

From a manager's perspective, this structure keeps governance concrete. Instead of abstract policies, you have specific checkpoints in real workflows.

Data Flow Governance — Consent, Access, and Purpose

FHIR makes it easier to move data. Governance ensures you move it **appropriately**.

At the data flow stage, you're answering three questions:

1. **Consent and purpose** – For what purposes can we use this data? Direct care? Operations? Research? Model training?
2. **Access control** – Who and which systems can see which FHIR resources, and at what level of detail?
3. **Data minimization** – How much data do we truly need for a given workflow or model?

Concrete practices:

- Defining which FHIR resources are in scope for which use cases. For example, you might allow broad use of Encounter and Observation data for care management AI, but tightly restrict access to certain sensitive Conditions.
- Scoping FHIR API permissions so that apps and AI services only access the resources they need, not the entire record.
- Documenting when data is used for **care delivery** versus **secondary purposes** (research, product development), and ensuring consent and policy cover that distinction.

As a manager, you don't need to design access control lists yourself. You do need to insist on three things:

- A **purpose-of-use** story for every FHIR data flow: "We're using this data to do X, for Y group, with Z safeguards."
- Visibility into which external partners (vendors, payers, apps) access your FHIR APIs and why.
- Clarity on who can approve new uses of FHIR data, especially for AI training.

When data flow is governed well, you can move faster because staff know what's allowed and what's not, instead of re-arguing boundaries on every project.

Insight Flow Governance — Fairness, Transparency, and Limits

In the insight flow stage, governance concerns how AI behaves with FHIR data—what it's allowed to do, how you check it, and how you explain it.

You're dealing with questions like:

- **Fairness:** Does this model's performance vary across patient groups in ways that conflict with our mission or obligations?
- **Transparency:** Can we explain, in plain language, why a patient was flagged or a claim was scored a certain way?
- **Limits:** For this use case, what error rates and false positives/negatives are acceptable? What's the maximum influence the model can have on decisions?

Practical moves include:

- Requiring a simple **model card** or summary for each AI system: purpose, inputs (FHIR resources), performance metrics, and known limitations.
- Running basic fairness checks where appropriate—comparing performance metrics across key demographic or clinical subgroups.
- Agreeing in advance on **risk tiers** for use cases (e.g., low, medium, high), with heavier governance requirements (more testing, more documentation) for higher tiers.

Your role is to make sure these checks are not theoretical. For each significant FHIR + AI workflow, you can ask:

- Who signed off on this model's use in this context?
- What population was it tested on, and how similar is that to ours?
- Where are its limitations documented, and how are users made aware?

If those answers are fuzzy, you have a gap in insight flow governance.

Action Flow Governance — Human Oversight and Accountability

Governance at the action stage is about **who is in charge** when AI meets human decision-making.

Key questions:

- When the AI suggests something, who is ultimately accountable for the decision?
- What is the expected role of the human—rubber stamp, active reviewer, or exception handler?
- How are overrides documented, and how easy is it to disagree when needed?

To make oversight real, you can:

- Define explicit **human-in-the-loop patterns** for different risk levels. For example:
    - Low-risk: AI drafts; human glances and approves most of the time.
    - Medium-risk: AI suggests; human must actively review and confirm.
    - High-risk: AI can only provide background or prompts; final decision must be fully human and documented.
- Provide simple **UI patterns** for override decisions—buttons or fields that capture “accepted,” “modified,” “rejected” along with a brief reason when appropriate.
- Make sure policies and training emphasize that clinicians and staff are empowered and expected to challenge the AI when it conflicts with their judgment or patient needs.

As a manager, you should be able to walk through a FHIR + AI-enabled workflow and answer:

- “If something goes wrong here, who is accountable?”
- “If a user thinks the AI is wrong, how do they push back?”
- “Do our policies and culture support that behavior?”

If people feel they must follow AI's suggestions to avoid blame, you have an action-flow governance failure waiting to happen.

Learning Flow Governance — Monitoring and Incident Response

Governance in the learning stage turns your symbiosis loop into a **continuous safety and improvement mechanism**, rather than a one-time launch.

You're trying to ensure:

- AI performance and fairness are monitored over time, not just at go-live.
- Drift, anomalies, and incidents are detected and handled systematically.
- Lessons learned from issues feed back into models, workflows, and policies.

Practically, this looks like:

- Defining **monitoring indicators** for each major FHIR + AI workflow: accuracy, utilization, overrides, potential bias indicators, and operational impact.
- Setting **thresholds** that trigger review e.g., if override rates spike, if performance drops below a certain level, or if certain groups see worse outcomes.
- Having a lightweight **incident playbook**: how staff reports suspected AI issues, how they are triaged, who investigates, and how changes are rolled out.

For each significant workflow, you should be comfortable asking:

- “What would count as a concerning signal here, and how would we notice?”
- “Who gets alerted, and what authority do they have to pause or adjust the system?”
- “How do we communicate back to staff about what happened and what changed?”

Without this, you risk quite a degradation. With it, you can treat small issues as early warnings instead of waiting for a headline-level problem.

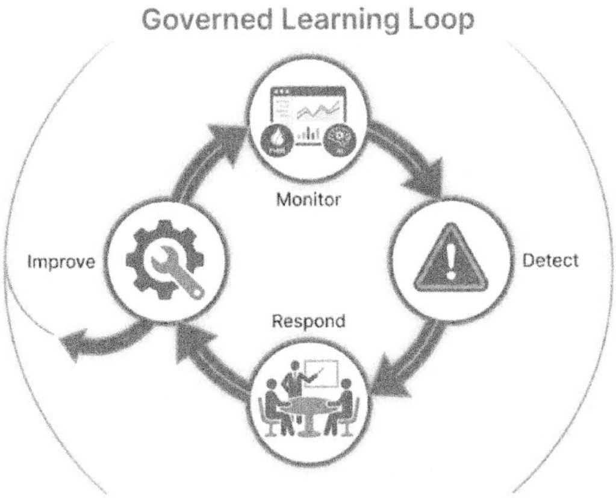

Practical Structures: Roles and Forums

Governance by design requires structure, but not necessarily more committees. You need clarity about **who does what** and **where decisions are made**.

Common elements:

- A cross-functional **AI/FHIR governance group** that:

  - Reviews higher-risk FHIR + AI use cases.
  - Approves risk tiers, guardrails, and monitoring plans.
  - Maintains a register of AI-enabled workflows and their owners.
- Named **product or workflow owners** for each major use case:
  - Accountable for value, risk, and continuous improvement.
  - Act as the focal point between clinicians, operations, IT, and compliance.
- Integration with existing structures:
  - Security and privacy forums handle FHIR access and incident review.
  - Quality and risk committees incorporate AI-related metrics and cases.
  - Change advisory boards include AI-specific considerations for go-live decisions.

As a manager, you don't have to create a heavy new governance layer. You do need to make sure AI and FHIR are visible in **the governance you already have**—not floating in a separate, informal space.

The Manager's Governance-by-Design Checklist

When a new FHIR + AI initiative is proposed—or when you review an existing one—use this checklist:

Purpose and Scope

- What problem are we solving, and for whom?
- Which FHIR resources and data elements are involved?

Data Flow

- Is data use aligned with consent and purpose-of-use policies?
- Are access controls scoped to what this workflow actually needs?

Insight Flow

- Which AI pattern(s) are we using?
- Do we have a brief model description (inputs, performance, limitations)?
- Have we considered basic fairness and transparency expectations?

Action Flow

- Who uses the outputs, where, and when?
- What is their responsibility relative to the AI?
- How are overrides and disagreements captured?

Learning Flow

- What metrics will tell us this is working—or not?
- How will staff report issues?
- Who can pause or adjust the system if something goes wrong?

If any of these answers are "we're not sure," you have a governance-by-design gap to close before scaling.

Chapter 6 Takeaways

- Governance by design is how you make FHIR + AI safe, trustworthy, and scalable—without turning innovation into chaos.
- It lives inside the symbiosis loop, shaping data flow, insight flow, action flow, and learning flow.
- Your job is to ensure that consent, access, fairness, transparency, human oversight, monitoring, and incident response are built into workflows—not added only after something goes wrong.

In the next chapter, we'll turn to **infrastructure**: how to build an AI-ready FHIR environment that supports these workflows at scale, across systems and vendors, without collapsing under its own complexity.

Chapter 7: Building the AI-Ready FHIR Infrastructure This chapter turns concepts into execution. It outlines how to build — or evolve — your architecture to make FHIR data usable for AI: integrating APIs, ensuring data quality, and layering analytic pipelines securely on top of clinical systems. It focuses on team structures, role clarity, and vendor collaboration, helping managers effectively bridge the gap between strategy and technical teams.

7 Building the AI-Ready FHIR Infrastructure

Why Infrastructure Is Now a Strategic Question

You already have systems: an EHR, billing, scheduling, care management tools, maybe a population health platform. You may even have some FHIR endpoints and AI pilots. But that does not mean you have an **AI-ready FHIR infrastructure**. Most organizations start with a patchwork of point-to-point feeds, vendor-specific APIs, and a few experimental models running on copied data. It works until you try to scale.

This chapter is about turning that patchwork into a **backbone**: an environment where FHIR is the standard way data moves, and AI is a standard way intelligence is applied. Infrastructure here is not just servers and networks. It's architecture, team structures, role clarity, and vendor integration patterns that let you go from one success story to a portfolio of reliable, governed AI-on-FHIR workflows.

Three-Layer Mental Model: Source, Backbone, Use

To keep things simple, use a three-layer mental model for your AI-ready FHIR infrastructure:

- **Source layer** – Where data originates (EHRs, ancillary systems, claims, external feeds).
- **Backbone layer** – Your FHIR environment plus data platform and governance.
- **Use layer** – The applications, AI services, analytics, and workflows that consume and write back data.

If you think only in terms of systems (EHR, data warehouse, AI tool), you miss the structure that connects them. The three layers give you a way to talk about architecture without having to draw wiring diagrams.

At the **source layer**, you're concerned with:

- Which systems hold critical data for your use cases?
- How data is extracted and normalized.
- How changes in source systems propagate.

At the **backbone layer**, you're concerned with:

- How FHIR resources are implemented and governed.
- How data quality is monitored and improved.
- How data is made available to AI and analytics securely.

At the **use layer**, you're concerned with:

- Where AI and analytics are embedded in workflows.
- How applications read from and write to the backbone.

- How performance, safety, and impact are monitored.

Your job is not to design every connector. It is to ensure each layer exists, is owned, and is connected by deliberate choices—not just historical accidents.

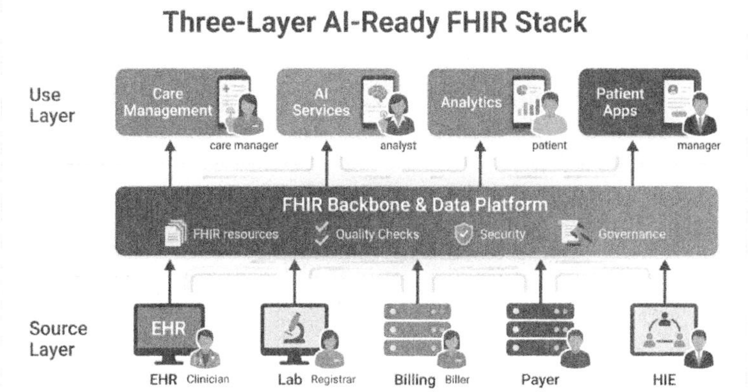

Designing the FHIR Backbone: Start with the Use Cases

An AI-ready FHIR infrastructure does not mean "implement all of FHIR." It means **implementing the parts of FHIR that matter most to your priority workflows, and doing so** well.

Start from use cases, not from the spec. For each flagship initiative (for example, readmission reduction, care gap closure, denial prevention), ask:

- Which FHIR resources are essential? (Patient, Encounter, Condition, Observation, MedicationRequest, Procedure, Coverage, etc.)
- What level of detail and completeness do we need? (All encounters or just inpatient? All labs or specific panels?)
- How quickly must data be available? (Near real time vs. daily vs. weekly.)

Then, work with architecture and informatics to:

- Define a **core resource set** you will treat as "Tier 1" for backbone investment.
- Align on **profiles** for those resources (e.g., US Core plus your internal extensions).
- Document **gaps** between today's reality and your desired state.

This gives your FHIR work a clear direction. Instead of a generalized "FHIR project," you have a backbone being built to support concrete AI-driven workflows.

The Data Platform: Where FHIR Meets Analytics and AI

FHIR endpoints are necessary but not sufficient. You also need a **data platform** that can:

- Ingest FHIR data from multiple sources.

- Store and curate that data in analysis-friendly formats.
- Serve it efficiently to AI and analytics workloads.
- Feed results back into operational systems when needed.

There are different technical approaches to this (data lakes, warehouses, lakehouses, cloud platforms), but the core idea remains constant: **FHIR is the contract at the edges**, while internally you may optimize storage for analytics and AI.

An AI-ready data platform typically:

- Ingests from FHIR APIs and other feeds into a consolidated environment.
- Maintains **lineage**—you can trace a feature or metric back to FHIR resources and then to source systems.
- Exposes curated datasets to AI teams through governed access patterns.
- Maintains **synchronization** with operational systems so models stay relevant.

From your seat, the key questions are:

- Do our AI and analytics teams have a stable, governed platform fed by FHIR, or are they constantly scraping data directly from source systems?
- Can we trace what data a model used, and when?

- Who owns the data platform's design, budget, and roadmap?

An AI-ready infrastructure is one where a new use case does *not* require a ground-up extraction project.

7.5 Security and Privacy: Building Trust into the Backbone

As you make data more available and introduce AI, your attack surface and risk profile change. Security and privacy cannot be an afterthought layered onto FHIR and AI. They need to be part of the backbone's design.

Practically, this means:

- **Identity and access management** aligned across systems—role-based access tied to clinical and operational roles, not just technical groups.
- **Scoped FHIR access** for each application and AI service: least privilege by default, explicit business justification for wider scopes.
- **Network and API protections**—rate limiting, anomaly detection, and strong authentication for external access.
- **Logging and auditability**—comprehensive logs for FHIR access, AI model calls, and data platform queries, with the ability to investigate suspicious patterns.

On the privacy side, you need clarity on:

- What constitutes **direct care** versus secondary data use.
- How data is **de-identified or pseudonymized** for model training, where possible.
- What consents or notices apply to FHIR-enabled data sharing and AI use?

As a manager, you're not configuring firewalls, but you are accountable for ensuring that:

- Security and privacy leaders are part of FHIR and AI design discussions, not just final sign-off.
- Infrastructure choices (e.g., cloud vs. on-premises, vendor APIs) match your risk appetite and regulatory environment.
- Breach and incident response plans explicitly cover FHIR APIs and AI systems.

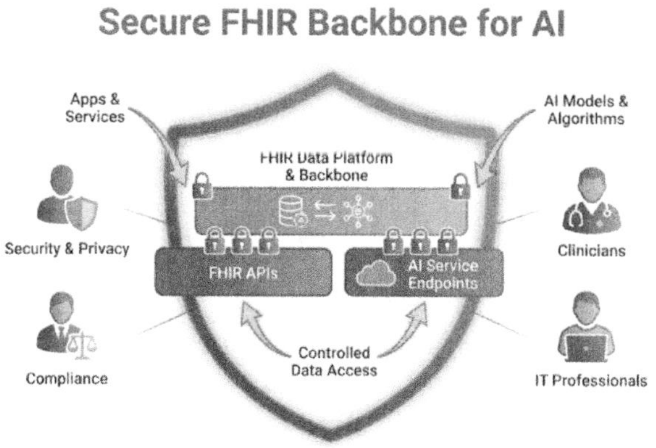

Team Structures: Who Owns What in an AI-Ready FHIR World

Infrastructure is as much about people and responsibilities as it is about technology. You need clear **ownership** to avoid stagnation and turf wars.

Key roles and groups typically include:

- **FHIR Product Owner / Architect**
  Owns the FHIR layer design and evolution. Decides which resources and profiles are supported, how they map to source systems, and how to manage change.
- **Data Platform Owner / Lead**
  Owns the analytics/AI platform fed by FHIR. Sets standards for ingestion, modeling, and data access; works closely with AI and analytics teams.
- **AI / Analytics Lead**
  Oversees AI use cases, model development practices, and integration with FHIR and applications.
- **Security and Privacy Leads**
  Define and enforce security and privacy controls for FHIR and AI, including third-party access.
- **Clinical and Operational Champions**
  Represent the workflows that will use AI on FHIR. Co-design how insights appear and how they are acted upon.

- **Governance Group**
  Provides cross-functional oversight for higher-risk use cases and major architectural decisions.

You do not need a large organizational chart. You do need to answer:

- Who signs off on changes to FHIR profiles that affect multiple projects?
- Who decides which AI models move from pilot to production?
- Who resolves conflicts between performance, cost, and risk?

When these responsibilities are implicit, infrastructure becomes fragile, and AI projects either stall or grow in unsupervised directions.

Working with Vendors: Plugging into Your Backbone, Not Replacing It

Vendors are part of your infrastructure, whether you like it or not. Many will come with their own AI capabilities, data models, and APIs. If you're not careful, you end up with **mini-FHIRs inside each product, each with its** own rules.

To build an AI-ready FHIR infrastructure, you need a **vendor posture**:

- Vendors should **use your FHIR interfaces** for data exchange where feasible, not demand separate one-off feeds for everything.
- Vendors that provide AI should be able to **consume FHIR data** and **return outputs** in ways that fit your symbiosis loop (e.g., as FHIR Observations/Flags, or via defined integration points).
- Contracts should reflect **data use and access** expectations explicitly: what data they can retain, how they can use it (or not) for their own models, and what happens at exit.

When evaluating or renegotiating:

- Ask vendors to show how their solution maps to your core FHIR resources and profiles.
- Require a clear description of how their AI models use your data and how they can be monitored.
- Clarify joint responsibilities for security incidents and AI misbehavior involving their products.

The message internally should be clear: "We have an AI-ready FHIR backbone. We buy tools that respect and strengthen that backbone, not ones that fragment it."

From Pilots to Platform: Reusing Infrastructure Patterns

Many organizations build their first FHIR + AI use cases as bespoke pilots. That's fine for learning—but it's not fine to stay there. The goal of an AI-ready FHIR infrastructure is to **turn one-offs into patterns**.

After each pilot, ask:

- What FHIR resources and profiles did we need? Can those be reused or generalized?
- What data pipelines and features did we build? Can they feed other models?
- What integration patterns did we use to surface AI outputs? Can those become standard UI components or APIs?

Over time, you want reusable assets such as:

- Standard FHIR-based patient risk views.
- Shared feature libraries for common predictors (e.g., recent utilization, chronic condition counts).
- Repeatable deployment pipelines for models using FHIR inputs.
- Shared logging and monitoring dashboards.

This is what shifts you from "we have some AI pilots" to "we have an AI platform on FHIR." New use cases then become **configuration and adaptation** problems rather than greenfield builds.

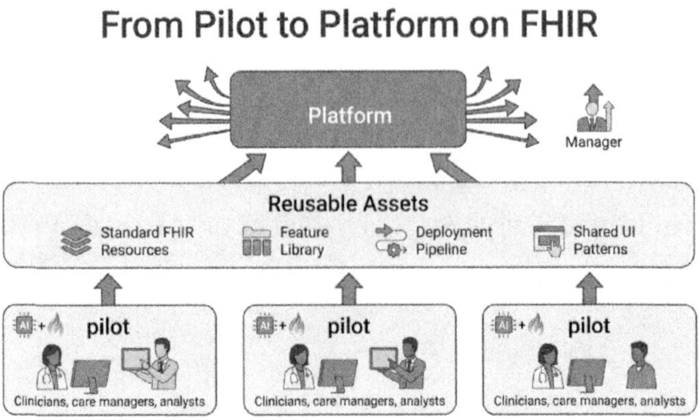

7.9 Practical Roadmap: Evolving to AI-Ready in 12–24 Months

You don't need to build a perfect infrastructure up front. You do need a **deliberate roadmap**. Over 12–24 months, a realistic progression might look like:

Phase 1: Baseline and Flagship Use Cases (0–6 months)

- Inventory current FHIR capabilities and AI/analytics efforts.
- Choose 1–3 flagship use cases where FHIR + AI can show value.
- For each, define required FHIR resources and data quality targets.
- Stand up minimal but solid FHIR and data platform paths for those use cases.

- Establish basic governance (ownership, risk tiering, monitoring expectations).

Phase 2: Consolidation and Pattern-Building (6–12 months)

- Harden FHIR implementations for core resources used in Phase 1.
- Package and document reusable data and integration patterns.
- Introduce more systematic logging, monitoring, and security controls around FHIR APIs and AI services.
- Pilot a small central AI/FHIR governance group focused on higher-risk workflows.

Phase 3: Expansion and Vendor Alignment (12–24 months)

- Extend FHIR coverage and data quality to additional high-value resources/use cases.
- Bring more internal and vendor AI tools onto the backbone, reducing siloed integrations.
- Formalize team structures and ownership for FHIR, data platform, and AI portfolios.
- Integrate FHIR + AI considerations into standard IT, security, and procurement processes.

You can move faster or slower depending on size, risk appetite, and resources. The key is that you're always moving in the **same direction**—toward a backbone that makes each new FHIR + AI initiative easier, safer, and more valuable than the last.

Chapter 7 Takeaways

- An AI-ready FHIR infrastructure is not “just” interfaces; it is a backbone plus platform plus governance, designed to support workflows at scale.
- Use the three-layer model (source, backbone, use) to structure conversations and decisions.
- Start your FHIR investments from high-value AI use cases, not from the spec; then reuse patterns across projects.
- Align security, privacy, team structures, and vendor relationships with your backbone, so FHIR and AI strengthen your operating model instead of fragmenting it.

In the next chapter, we’ll focus on **execution over time**: how to move from pilot experiments to sustainable, production-grade FHIR + AI capabilities that you can trust, measure, and expand.

8 From Pilot to Production — Making FHIR + AI Stick

AI-FHIR pilots are easy to start but hard to scale. This chapter provides a blueprint for progress: defining scope, securing data sources, measuring early outcomes, and developing a sustainability plan. Managers learn the "Pilot-to-Production Pathway" — a repeatable process that turns experiments into enterprise capabilities aligned with mission and regulatory readiness.

Why Pilots Succeed and Programs Stall

Most organizations can point to at least one "successful" FHIR + AI pilot—a unit where readmissions dropped. A documentation assistant that clinicians liked. A care gap dashboard that impressed leadership. The pattern is familiar: the pilot team is hand-picked, the data is curated, the vendor or internal team is highly engaged, and everyone is watching.

Then comes the hard part: scaling beyond that protected bubble. When you try to deploy the same capability to other units and teams under normal conditions, the shine fades. Data quirks appear. Workflows differ. Adoption is uneven. Monitoring and support feel heavier. Governance questions multiply. What worked as a one-off project doesn't translate into a repeatable capability.

This chapter is about closing that gap. You'll learn a **Pilot-to-Production Pathway**—a simple, repeatable way to move from idea to stable, enterprise-grade FHIR + AI workflows that your teams can actually live with.

The Pilot-to-Production Pathway

You can think of the pathway in four stages:

1. **Explore** – Clarify the problem, value, and feasibility.
2. **Prove** – Run a focused pilot under realistic conditions.
3. **Harden** – Strengthen data, workflows, and guardrails for broader use.
4. **Scale** – Roll out systematically, monitor, and continuously improve.

Each stage has a different goal, a different definition of success, and different governance expectations. Treating them as distinct helps you avoid two common mistakes:

- Declaring victory after a pilot and "throwing it over the wall" to operations.
- Holding pilots to full production standards before you've even established basic value.

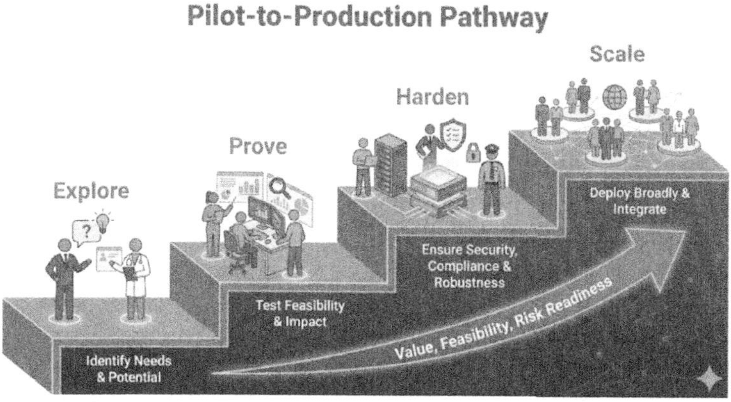

Stage 1 — Explore: Define the Problem and the Loop

In the **Explore** stage, your goal is not to build anything yet. Your goal is to define:

- The **problem** you're trying to solve.
- The **symbiosis loop** for this use case (data, insight, action, learning).
- The **value hypothesis** and basic risks.

You should be able to write a short narrative in plain language:

- "We believe that using FHIR + AI to prioritize discharge planning on the medicine unit will reduce 30-day readmissions by 10 percent over six months while not increasing length of stay."

And sketch, at a high level:

- **Data Flow** – Which FHIR resources are needed (Patient, Encounter, Condition, Observation, MedicationRequest, Coverage, etc.).
- **Insight Flow** – Which AI pattern dominates (likely prediction + prioritization).
- **Action Flow** – Who will act (care managers, hospitalists), where (huddles, EHR views), and when (daily).
- **Learning Flow** – How you'll measure impact (readmission rate, override patterns, staff feedback).

You also surface constraints:

- Data gaps ("We don't currently capture X reliably").
- Workflow constraints ("Care managers are already overloaded at these times").
- Governance concerns ("We need to be cautious about bias across certain subgroups").

If you can't articulate this clearly, you're not ready to move on. A common failure pattern is to skip Explore and jump straight to building a model because the data is available.

Stage 2 — Prove: Pilot Under Real Conditions

In the **Prove** stage, you run a pilot. The goal is to test whether the FHIR + AI workflow delivers value under realistic, not perfect, conditions, with guardrails appropriate to the risk.

Characteristics of a good pilot:

- **Clear scope** – One or a few units, a defined patient segment, limited time frame (e.g., 3–6 months).
- **Representative conditions** – Use the actual FHIR feeds, systems, and staff that would be involved in production, not a shadow environment with extra support that won't exist later.
- **Defined metrics** – A small set of outcome, process, and safety metrics you'll track (e.g., readmission rate, time spent per discharge, override rate, basic fairness checks).
- **Explicit guardrails** – A simple model description, documented owners, basic monitoring, and a rollback plan if something goes wrong.

During the pilot, you're asking:

- Does the workflow **fit** how people actually work?
- Are the **data quality issues** manageable or show-stoppers?
- Do the **metrics** show enough improvement to justify more investment?
- What does frontline feedback tell you about usability, trust, and unintended consequences?

The output of Prove is not just "it worked" or "it didn't." It's a decision: **should we harden and scale, adjust, or stop?** You capture lessons learned about data, workflow, and governance that you'll need later.

Stage 3 — Harden: Make It Real-World Tough

Assuming the pilot shows promise, the **Harden** stage is where you make the solution robust enough for broader exposure. This is where many organizations falter—they try to scale a pilot that still depends on heroic effort and fragile plumbing.

Hardening focuses on:

- **Data robustness** – Strengthening FHIR implementations and data quality for the resources the workflow relies on. Reducing manual workarounds. Ensuring performance is adequate as volume grows.
- **Operational integration** – Embedding the workflow into standard operating procedures, huddles, and training. Removing temporary pilot exceptions (e.g., extra staffing, special attention from experts).
- **Technical reliability** – Ensuring APIs, data platform, and model services are monitored, backed up, and supported like other critical systems. Building deployment and rollback mechanisms.
- **Governance maturity** – Finalizing risk tiering, documenting model behavior and limitations, agreeing on monitoring thresholds, and clarifying who can pause or update the system.

Some changes made in pilots may be reversed here. For instance, a pilot might have a model running nightly with manual checks; hardening might require automating parts of that process or tightening integration into existing IT practices.

A hardened solution can **survive attention**: more users, more data, more scrutiny from leadership and regulators.

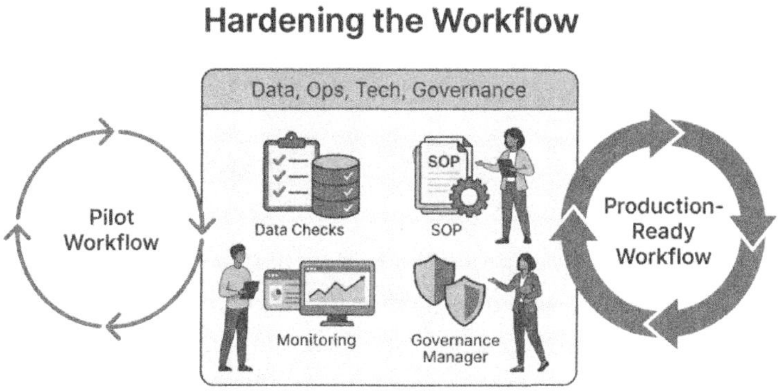

Stage 4 — Scale: Roll Out Deliberately

In the **Scale** stage, you broaden adoption. The goal is not just to "turn it on everywhere," but to **spread and adapt** the hardened pattern in a controlled way.

Key elements:

- **Sequenced rollout** – Decide which units or populations go next, based on readiness and value. You might start with units similar to the pilot, then move on to more complex ones.

- **Standardized training and change support** – Update training materials, job aids, and communication plans from the pilot into reusable assets. Identify local champions in each new area.
- **Monitoring at scale** – Ensure dashboards and alerting can handle increased volume. Track performance and adoption across sites, comparing to pilot expectations.
- **Feedback and iteration** – Treat each new rollout as another learning opportunity; collect feedback, adjust thresholds or UI patterns, and update playbooks.

Scaling is also where you handle **variation**. Different units may have different workflows or constraints. The art is in maintaining a consistent core (data, model, guardrails) while allowing limited, intentional customization to fit local realities.

At this stage, leadership asks: "Can we trust this across the enterprise?" Your answer depends on how well you did in Harden and how disciplined you are in Scale.

The Manager's Role at Each Stage

Your responsibilities evolve across the pathway.

In **Explore**, you:

- Frame the problem in operational and mission terms.
- Bring together clinical, operations, IT, and data voices to define the symbiosis loop.

- Decide whether the idea is worth piloting at all.

In **Prove**, you:

- Secure resources and sponsorship for a pilot.
- Ensure the pilot uses realistic data and workflows, not artificial conditions.
- Keep focus on the defined metrics and feedback, not just anecdotes.

In **Harden**, you:

- Push teams to address the unglamorous work: data quality, support, documentation, SOP updates.
- Involve governance, security, and compliance properly.
- Resist pressure to scale prematurely before robustness is proven.

In **Scale**, you:

- Prioritize rollout sequence based on readiness and impact.
- Coordinate change management across multiple sites or teams.
- Use monitoring and feedback to steer adjustments and decide where to invest next.

Seeing your role as changing across stages helps avoid both micromanagement ("I must be in every technical decision") and abdication ("IT will handle it"). You're steering the **pathway**, not the code.

Common Pitfalls Between Pilot and Production

Several predictable traps show up when moving from pilot to production:

- **Pilot glow** – Declaring success too soon based on hand-picked users and heavy support that won't exist later.
- **Data shortcuts** – Using one-off, curated datasets for pilots and then discovering that real FHIR feeds are messier and slower.
- **Workflow mismatch** – Designing pilots around exceptionally engaged teams whose practices don't match the rest of the organization.
- **Governance lag** – Launching pilots under informal approvals and only trying to formalize governance when scaling, creating friction and rework.
- **Technical fragility** – Building pilots on "temporary" infrastructure that cannot meet reliability or security expectations at scale.

You can mitigate these by:

- Insisting pilots use the **same FHIR backbone** you intend to run in production, even if the scope is smaller.
- Selecting pilot sites that are slightly above average in readiness, not just the most innovative outliers.
- Engaging governance and security early enough that they help shape the design, not just approve it.
- Treating infrastructure decisions in pilots as **first drafts** of your platform, not throwaways.

Recognizing these pitfalls ahead of time lets you design pilots with an eye toward production, not just proof.

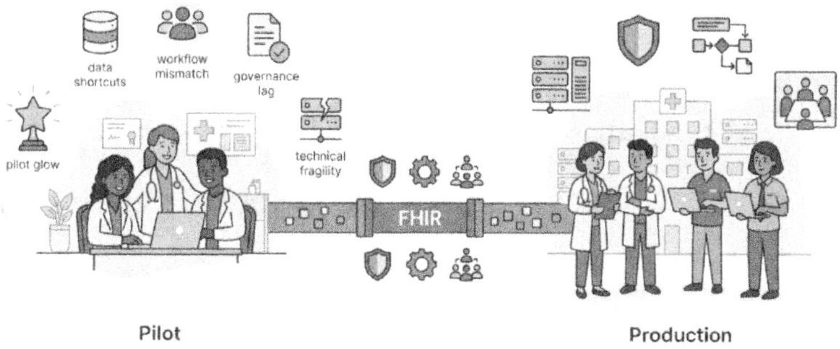

The Pilot-to-Production Canvas

To make the pathway operational, you can use a simple one-page canvas for each FHIR + AI initiative. It has four sections—Explore, Prove, Harden, Scale—each with a few prompts.

For example:

- **Explore**

    - Problem statement
    - Target population and workflow
    - FHIR resources needed
    - Primary and secondary metrics
    - Risks and assumptions
- **Prove**
    - Pilot scope (where, who, how long)
    - Data sources and environment
    - Minimum guardrails (ownership, monitoring, rollback)
    - Success criteria
- **Harden**
    - Data quality and FHIR enhancements required
    - Operational integration (SOPs, training, support)
    - Technical hardening (performance, resilience)
    - Governance decisions (risk tier, approvals, documentation)
- **Scale**
    - Rollout plan (sequence, dependencies)
    - Change management approach
    - Monitoring plan and thresholds

- Improvement feedback loop

You don't need a consulting-grade template. You need something teams actually fill out and revisit.

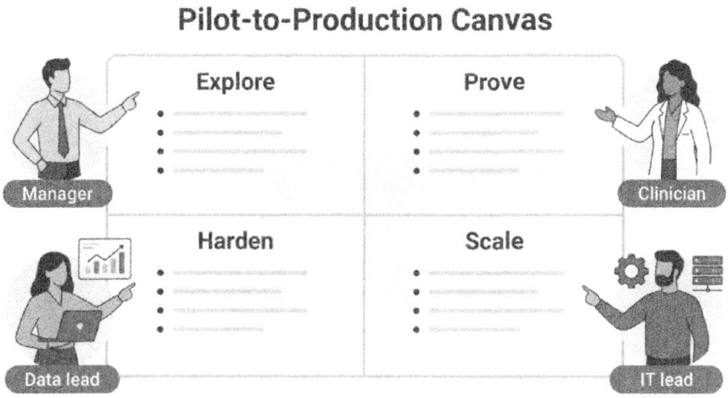

Chapter 8 Takeaways

- Pilots are easy; production is hard. You need a deliberate, repeatable pathway to turn FHIR + AI experiments into durable capabilities.
- The Pilot-to-Production Pathway (Explore, Prove, Harden, Scale) helps you align expectations, governance, and investment at each stage.
- Designing pilots on your real FHIR backbone, with realistic workflows and early governance, is the single best way to avoid painful rework later.
- As a manager, you steer the pathway, ensuring that each stage produces the evidence and hardening needed for the next.

In the next chapter, we'll deepen the **Responsible AI Lens**—showing how fairness, transparency, and accountability run through every stage of the pathway, not as extra steps but as integral parts of your operating model.

9 The Responsible AI Lens in Healthcare

This chapter brings an ethical and operational focus to AI-driven decision-making. It covers fairness, bias detection, outcome transparency, and explainability. Readers learn how to apply a "Responsible AI Lens" throughout each project stage — from dataset selection to post-deployment monitoring. The key idea: responsibility is not a phase; it's a posture that protects both patients and the organization

Why Responsibility Cannot Be Bolted On

An AI model in healthcare is never just a piece of code. It sits between a patient and a clinician, between a claim and a payment, between a member and access to services. When it goes wrong, the impact is not abstract; it hits real people and your organization's credibility.

Many organizations still treat "responsible AI" as a review step at the end: a checklist before go-live, a presentation to a committee, a legal review of a vendor contract. That mindset fails in practice. By the time you reach the end, your data choices are locked in, your model is trained, and your workflow is built. Fixes are expensive, partial, or purely cosmetic.

In healthcare, responsibility must be a **posture**, not a phase. It has to show up in how you choose problems, what data you collect, how you train and evaluate models, how you embed them in workflows, and how you monitor them over time. This chapter gives you a simple **Responsible AI Lens** you can apply at every stage of your FHIR + AI lifecycle, without needing to be a data scientist or ethicist.

The Responsible AI Lens: Four Questions

You can express the Responsible AI Lens as four recurring questions:

1. **Value** – Who is this helping, and how do we know?
2. **Fairness** – Who could be harmed or systematically disadvantaged?
3. **Transparency** – Who needs to understand what the AI is doing, at what level of detail?
4. **Accountability** – Who owns outcomes and decisions, and how do we learn from mistakes?

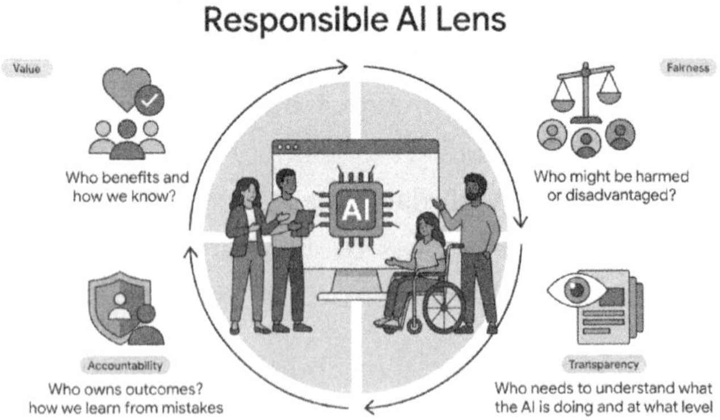

These questions are not academic. You can ask them in plain language in design meetings, pilot reviews, and governance forums. They help you move from vague comfort ("this seems fine") to concrete assurance ("we have thought about and tested for the right things").

Responsibility doesn't mean saying no to AI. It means saying **yes** with your eyes open, and with structures in place to catch issues early.

Fairness in Healthcare AI: Practical Concerns

Fairness in healthcare is not a theoretical slogan. It shows up when:

- A triage tool systematically under-prioritizes patients from certain neighborhoods.
- A readmission model performs worse for people with rare conditions.
- A care gap system focuses outreach on populations more likely to respond, leaving others further behind.
- A documentation assistant occasionally omits important context more often in one language than another.

In practice, fairness problems often come from:

- **Biased data** – Historical records that reflect unequal access, underdiagnosis, or different documentation practices.
- **Incomplete data** – Missing fields (e.g., race, language, social risk) that hide disparities.
- **Proxy features** – Variables that unintentionally stand in for protected attributes (e.g., ZIP codes, certain utilization patterns).
- **Optimization choices** – Focusing solely on overall accuracy or savings, without looking at who benefits and who doesn't.

A Responsible AI Lens does not demand perfection. It does demand that you:

- Identify which groups you care about from a fairness perspective (e.g., race, ethnicity, language, age, disability, payer type, geography).
- Compare model performance and outcomes across those groups where data and law allow.
- Decide, explicitly, what disparities are acceptable and what mitigation steps you will take if you find more than that.

You don't need to design fairness metrics yourself. You do need to make fairness **visible** and **to discuss it** when decisions are made.

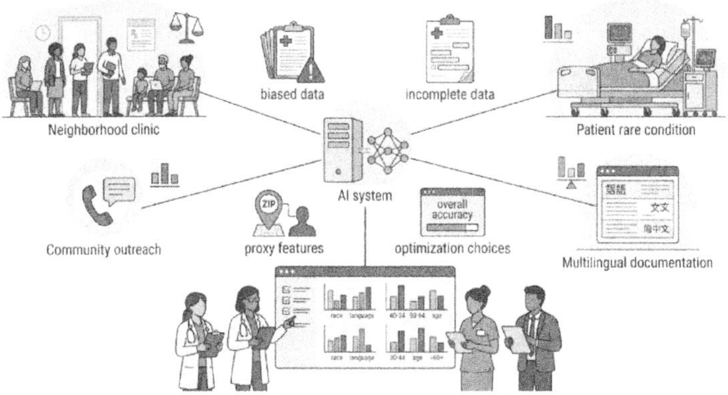

Bias Detection and Mitigation Along the Lifecycle

Bias detection is not a one-time test. It needs to appear in multiple stages of your AI-on-FHIR lifecycle:

- **Dataset selection and preparation**
  - Is the data representative of the population you will deploy to?
  - Are key groups under-represented or missing?
  - Are important outcomes systematically under-documented for certain groups?
- **Model training and evaluation**
  - When evaluating performance, do you break results down by key subgroups?
  - Do you look at false positive/negative rates, not just overall accuracy or AUC?

- Are there obvious disparities that require a design change?

- **Pilot and deployment**
  - During pilots, do you monitor usage and outcomes across groups (e.g., who gets flagged, who gets outreach, who improves)?
  - Do pilot sites include diverse settings, not just "easy" ones?
- **Post-deployment monitoring**
  - Are fairness checks part of your regular monitoring dashboards, alongside performance and volume?
  - Do you revisit fairness when your population changes, when you expand to new sites, or when external conditions shift?

Mitigation can take several forms:

- Adjusting training data (e.g., re-balancing samples, improving data quality for under-represented groups).
- Changing model features or targets.
- Introducing policy rules on top of model outputs (e.g., minimum attention to certain populations).
- Tightening where and how the model is used (e.g., advisory only in higher-risk scenarios).

The point is not to eliminate all bias—that's impossible—but to **understand, reduce, and govern** it deliberately.

9.5 Outcome Transparency: Being Able to Show Your Work

In healthcare, you will be asked "why":

- "Why was this patient flagged as low risk?"
- "Why was my claim flagged for extra review?"
- "Why did I not receive outreach when others did?"
- "Why is this model allowed to influence care?"

Outcome transparency is your ability to answer those questions with more than "the model said so."

Practically, this means:

- Documenting, in simple language, each model's purpose, inputs (which FHIR resources and key fields), and main risk drivers.
- Ensuring user interfaces show **key contributing factors** for individual decisions or scores when appropriate (e.g., past admissions, recent labs, specific conditions).
- Making clear where AI is involved in a workflow and what role it plays (advisory vs. driving a queue vs. gating access).

Transparency is not all-or-nothing. Different audiences need different levels:

- Clinicians may need to see which clinical factors drove a score.
- Patients may need a plain-language explanation of how tools support care decisions.
- Executives and regulators may need a higher-level view of models in use, their performance, and governance.

A Responsible AI Lens asks, at design time: **who will ask us "why" here, and what will we show them?**

Explainability: Glass Box Enough for the Task

Explainability is related to transparency, but narrower: it's about how well you can understand a model's inner workings and why it produced a specific output.

Not every AI model in healthcare needs the same level of explainability. For example:

- A model that suggests wording for low-stakes patient reminders can be less interpretable, as long as guardrails prevent harmful content.
- A model that influences triage or treatment decisions should be much more explainable, or at least operate under stricter human oversight with clear constraints.

Explainability options include:

- Using models that are inherently interpretable when stakes are high (e.g., simpler models with clear feature contributions).
- Providing post-hoc explanations (e.g., feature importance lists, local explanations) where appropriate.
- Avoiding model designs or integrations where you cannot explain behavior at a level that satisfies your clinicians, patients, and regulators.

The Responsible AI Lens here is simple: **for this use, is our level of explainability adequate for the risk?** If you can't convincingly answer "yes," you either need more explainability or a different model or workflow design.

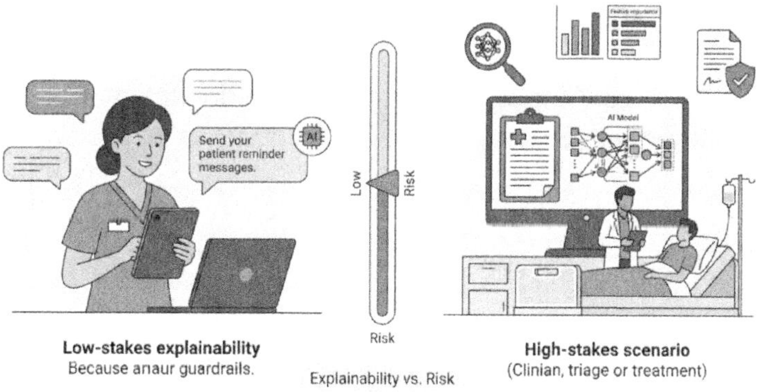

Responsibility Across the Project Stages

Responsibility looks different at each stage of your Pilot-to-Production Pathway:

- **Explore**
    - Ask: Is this problem worth solving with AI? Are there simpler, lower-risk options?
    - Identify stakeholders who could be harmed if this goes wrong (patients, specific populations, staff, payers).
- **Prove**
    - Define minimum fairness and transparency expectations up front.
    - Collect qualitative feedback on trust and usability from clinicians and staff.
    - Ensure pilots include diverse populations or make explicit that fairness conclusions are limited.
- **Harden**
    - Formalize fairness tests, documentation, and monitoring plans.
    - Confirm that explainability and transparency are adequate for the expanded scope.
    - Integrate responsible AI checkpoints into SOPs and governance processes.
- **Scale**
    - Continuously monitor for drift and emerging disparities as volume and diversity increase.

- Update communications and training to reflect what you've learned about limitations and proper use.
- Build incident response pathways that explicitly cover ethical and fairness concerns, not just technical outages.

Responsibility is not something you "add" at the end. It's a thread you weave through decisions from day one.

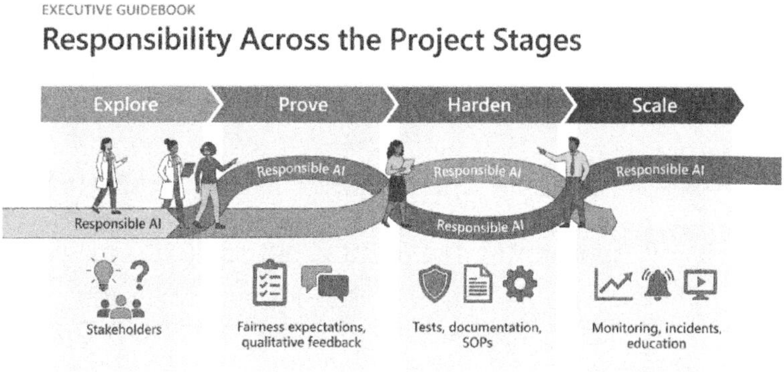

Governance Structures That Make Responsibility Real

A Responsible AI Lens needs a home in your governance structures, not just in slide decks. That does not always mean building a new committee, but it does mean:

- Ensuring your AI/FHIR governance group explicitly considers fairness, transparency, and accountability for higher-risk workflows.

- Giving clinical leaders and, where appropriate, patient or community representatives a voice in high-impact decisions.
- Making responsible AI expectations visible in policies, templates, and checklists (e.g., requiring basic subgroup analyses for certain use cases).

You also need clear **owners** for:

- Responsible AI practices (someone who keeps this lens from disappearing under pressure).
- Communication to clinicians, staff, and, where relevant, patients about AI's role and limitations.
- Responding when responsible AI concerns are raised.

Responsibility is a posture, but posture is maintained by muscles—your structures, roles, and routines.

The Manager's Responsible AI Playbook

Here is a compact set of questions you can integrate into your existing decision processes. Use these in intake meetings, design reviews, pilot debriefs, and governance forums:

- **Value**
  - What outcome, for whom, are we trying to improve?

    - How will we measure whether this is happening, and over what time frame?
- **Fairness**
    - Which groups could be affected differently by this system?
    - How will we check for disparities in performance or outcomes across those groups?
    - What will we do if we find concerning differences?
- **Transparency & Explainability**
    - Who needs to know that AI is involved here?
    - What explanation will clinicians, staff, and patients see, if any?
    - Is that explanation meaningful at their level of expertise?
- **Accountability & Monitoring**
    - Who owns this system in production?
    - How will we detect problems (performance drops, drift, bias, misuse)?
    - Who can pause or adjust the system, and how do we communicate changes?

You can start applying this playbook immediately, even to existing workflows. It will surface gaps you can address over time.

Chapter 9 Takeaways

- In healthcare, AI decisions are never neutral; they affect access, safety, equity, and trust.
- A Responsible AI Lens—focused on value, fairness, transparency, and accountability—must be applied throughout the FHIR + AI lifecycle, not just at the end.
- Fairness and explainability are practical concerns: you don't need perfection, but you do need visibility, monitoring, and clear thresholds for action.
- Responsibility is a posture woven into how you select use cases, handle data, design workflows, and govern systems—not a separate project.

In the next chapter, you'll see how this lens meets a very specific concern: **Securing FHIR**—protecting the backbone that all of your AI and interoperability efforts depend on, so trust in your data environment matches the ambition of your AI ambitions.

10 Securing FHIR — Protecting the Backbone of Interoperability

When Interoperability Becomes an Attack Surface

A large health system finally gets its FHIR program moving. Patient apps can pull records. Vendors integrate more easily. AI pilots use standardized data rather than scraping the EHR. For a while, everything feels like progress.

Then security starts asking harder questions. A third-party app requests broad access scopes "to support future features." An internal AI team spins up a cloud service to pull FHIR data from the network perimeter. An auditor wants to know exactly which users and vendors have accessed which FHIR endpoints for which patients. No one can answer quickly. The same openness that made FHIR powerful has quietly expanded the attack surface.

Security is not a separate concern bolted onto FHIR and AI. It is the **hinge between trust and use**. If you cannot defend your FHIR backbone, you will hesitate to expose it, limit AI to small experiments, and retreat to brittle point-to-point integrations. This chapter shows you how to secure FHIR in a way that **enables** interoperability and AI, rather than shutting them down.

FHIR Backbone Under Shield

What Makes FHIR Security Different

Securing FHIR is not the same as securing legacy interfaces. A few characteristics change the game:

- **Fine-grained, patient-level APIs** – Instead of nightly file drops, FHIR exposes records via HTTP calls that can be made at scale, with flexible queries.
- **Wider ecosystem access** – Patient apps, vendors, payers, and internal services may all call your FHIR endpoints.
- **Cloud and hybrid patterns** – FHIR servers and AI services often run in cloud environments, changing where and how controls apply.
- **Higher visibility and expectations** – Regulators, partners, and patients increasingly know what FHIR is; they expect you to manage its security intentionally.

These differences mean traditional "perimeter-only" thinking is not enough. You need to control **who can**

access what FHIR data, for which purposes, from where, and under what conditions, and you need evidence that you're doing so.

For you as a manager, this is not about configuring TLS ciphers. It is about ensuring your FHIR program includes security patterns that align with the standard's openness and granularity.

A Manager's FHIR Security Model: Five Control Layers

You can think about FHIR security through five control layers:

1. **Identity and access** – Who (people and systems) can access FHIR, and how they authenticate.
2. **Authorization and scopes** – What those identities are allowed to do and see.
3. **Transport and storage protection** – How data is protected in transit and at rest.
4. **Monitoring and detection** – How you see and respond to suspicious activity.
5. **Governance and contracts** – How policies and agreements align with technical controls.

These layers apply whether your FHIR server is in your data center, in the cloud, or provided by a vendor. Your job is to ensure that all five are **covered coherently**, not left to ad hoc decisions on a project-by-project basis.

Five Layers of FHIR Security

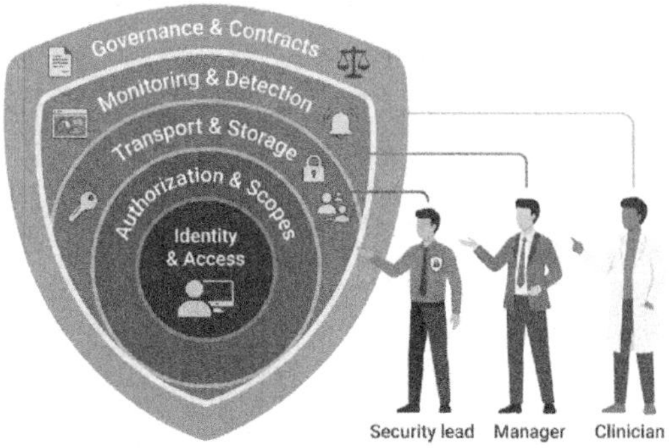

Identity and Access: Knowing Who's Knocking

Every FHIR call should come from a **known, authenticated identity**—either a user or a system. Weak identity is the starting point of most breaches.

Key expectations:

- **Centralized identity** – Clinicians, staff, and internal services use enterprise identity systems (e.g., SSO, directory) rather than ad hoc accounts for each FHIR app.
- **Strong authentication** – Multifactor for higher-risk roles and remote access; certificate-based authentication or secure service identities for system-to-system calls.
- **App registration** – Third-party apps (internal and external) that call FHIR APIs are registered, with owners, contact info, and approved scopes.

As a manager, you can ask:

- Do our FHIR servers rely on the same identity and SSO platform we use for our core apps?
- Can we list the apps and services currently authorized to call our FHIR APIs, with owners and purpose?
- Are high-privilege accounts and service identities managed with appropriate additional controls?

If the answers are unclear, your identity layer is not ready for serious exposure to FHIR.

Authorization and Scopes: Least Privilege on FHIR

Once you know who is calling, you must control **what** they can see and do. FHIR's fine-grained nature is both a risk and an opportunity here.

Practical patterns:

- **Role-based access control** – Map clinical and operational roles to sets of FHIR permissions (e.g., a nurse on unit X can read and write certain resources for patients on that unit; an analyst might have read-only access to a limited dataset).
- **Scopes for apps and services** – Third-party apps receive specific FHIR scopes (e.g., read-only, specific resource types, patient-level access) instead of blanket "all" privileges.

- **Purpose limitation** – Internal AI and analytics services are granted access tied to specific use cases, not open-ended access "for any future model."

Manager-level questions:

- When a new vendor tool needs FHIR, who decides which resources and scopes it receives?
- Do we have different profiles and scopes for patient-facing apps, partner apps, internal AI, and admin tools?
- Are there **guardrails** to prevent someone from accidentally granting overbroad access in a rush?

The default should be **least privilege**: start with the minimal access needed, then expand deliberately when justified.

Protecting Data in Transit and at Rest

FHIR traffic should be treated like any other sensitive healthcare data, with consistent protections:

- **In transit** – All FHIR endpoints use strong encryption (HTTPS/TLS). External access is limited to known networks and patterns. No unencrypted internal shortcuts for convenience.
- **At rest** – FHIR servers and data platforms store data with encryption at rest, with controlled key management. Backups follow the same rules.

- **Segmentation** – FHIR environments used for testing, development, and production are separated, with realistic but anonymized or de-identified data where appropriate.

You don't need to specify algorithms. You do need to insist that:

- FHIR environments follow your standard security baseline, not a weaker "innovation" baseline.
- Cloud-based FHIR or AI services meet your encryption and key management expectations.
- Non-production environments don't quietly become sources of data leakage because controls are lax "for speed."

If a security review reveals a separate, lightly controlled FHIR stack spun up for a pilot, that is a risk you must either bring into compliance or shut down.

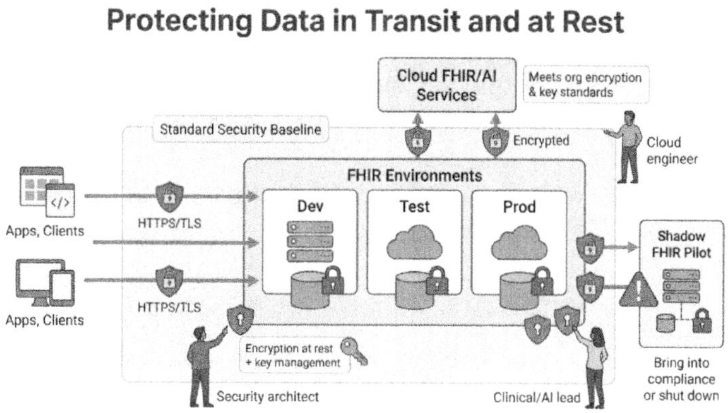

Monitoring, Logging, and Threat Detection

Securing FHIR is not just about blocking bad access; it is about **seeing** what's happening and reacting quickly.

Operational expectations:

- **Comprehensive logging** – FHIR servers log who accessed what, when, from where, using which app or account, and whether access was granted or denied.
- **Integration into SIEM/SOC** – Those logs feed into your broader security monitoring (SIEM, SOC), so patterns can be analyzed alongside other signals.
- **Anomaly detection** – Rules or models flag unusual patterns: high-volume access from unexpected locations, repeated attempts to access restricted resources, or apps suddenly requesting more data than usual.

From a manager's perspective:

- Can we quickly answer "which apps and users accessed patient X's FHIR data last month?"
- Do we have defined thresholds and alerting rules for suspicious FHIR usage?
- Who is responsible for investigating FHIR-related security alerts, and how do they coordinate with clinical and operations leaders when something serious is suspected?

Without monitoring, you will discover FHIR-related issues only when someone else brings them to you—a regulator, a partner, or the press.

Watching the FHIR Gate

Securing FHIR in Cloud and Hybrid Environments

Many AI-ready FHIR implementations live partly or entirely in the cloud. That changes some technical details, but it's not your responsibility.

Key cloud considerations:

- **Shared responsibility** – Cloud providers secure the infrastructure; you remain responsible for configuring FHIR servers, IAM, network rules, and data governance correctly.

- **Network boundaries** – Use private connectivity, VPNs, or VPC peering rather than wide-open internet exposure. Restrict which networks can reach FHIR endpoints.
- **Configuration drift** – Treat FHIR infrastructure as code where possible, so security settings are consistent across environments and changes are tracked.

Multi-cloud or hybrid setups add complexity: different providers, different tools, more places where misconfiguration can slip in.

Your questions:

- For each cloud-hosted FHIR or AI service, do we have a simple **responsibility map**—what the vendor secures, what we secure?
- Are we applying the same identity, access, and logging standards to cloud FHIR as we do on-premises?
- How do we detect and correct misconfigurations introduced by hurried changes or new projects?

Cloud can improve security when used well—but only if you bring your governance and expectations with you.

Third Parties and Contracts: Security Beyond Your Walls

Vendors, partners, and external apps often access your FHIR backbone. Contracts and due diligence are where you translate your security posture into enforceable obligations.

You will want to see:

- **Clear data use clauses** – What FHIR data the vendor can access, for what purposes, and whether they can retain or use it to train their own models.
- **Security commitments** – Baseline expectations (encryption, access controls, logging, incident response) aligned with your frameworks (HIPAA, HITRUST, NIST, internal policies).
- **Right to audit or assess** – The ability to receive security attestations, audit reports, or conduct assessments where warranted.
- **Incident obligations** – Timelines and content for breach notification, cooperation in investigations, and remediation responsibilities.

For AI-heavy vendors, also consider:

- How they secure **model training environments** that may ingest your FHIR data.
- Whether they support **tenant isolation** and strong access controls in SaaS offerings.
- How they will help you answer questions from regulators or patients about their use of your data.

Every contract is a chance to reinforce your architecture: "We operate an AI-ready FHIR backbone.

Solutions must plug into this backbone securely and respect our data and security standards."

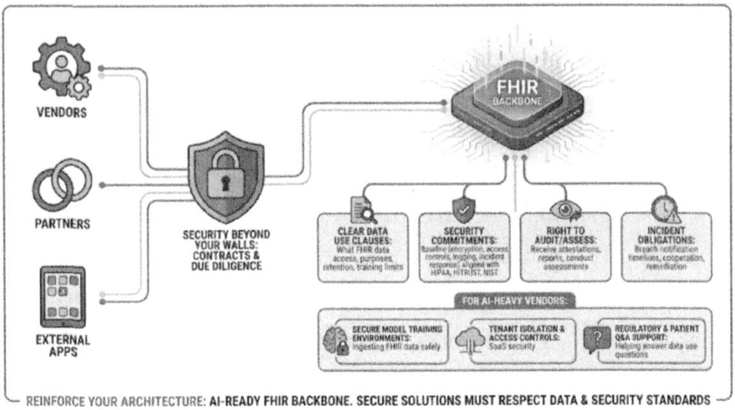

Security as Part of Your FHIR and AI Operating Model

Securing FHIR is not a one-time project. It is part of your **operating model** for FHIR and AI:

- New FHIR endpoints and resources are reviewed for security implications, not just data and functionality.
- New AI and app use cases must specify their FHIR access needs and demonstrate alignment with security and privacy expectations.
- Security incidents, near-misses, and red-team findings feed back into your FHIR and AI design patterns, improving them over time.

Practically, this means:

- Including security and privacy leads in your AI/FHIR governance group.
- Adding simple security questions to your intake templates ("What FHIR scopes are needed?", "External access?", "Cloud components?").
- Reviewing your FHIR security posture periodically as part of broader risk reviews.

You want security to be **predictable**: teams know what's required, security knows how to engage, and leadership sees FHIR and AI as safer because guardrails are clear.

The Manager's FHIR Security Checklist

To keep this actionable, use this checklist when reviewing FHIR-enabled projects or vendors:

- **Identity & Access**
    - Are all users and apps calling FHIR APIs authenticated via enterprise-approved mechanisms?
    - Do we know who owns each app or service account?
- **Authorization & Scopes**
    - Are FHIR permissions scoped to the minimum needed for this use case?

    - Who approved those scopes, and how are changes handled?

- **Protection in Transit & At Rest**
    - Does this solution follow our encryption standards for FHIR data?
    - Are non-production environments handled safely?
- **Monitoring & Detection**
    - Are FHIR access and AI model calls logged and integrated into security monitoring?
    - Do we have alerts for unusual access patterns?
- **Third Parties & Cloud**
    - Are vendor and cloud responsibilities clearly defined and documented?
    - Do contracts reflect our data use and security expectations?

You don't need to answer every question personally. You do need to insist that someone does so, and that the answers are documented, before you expose more of your backbone.

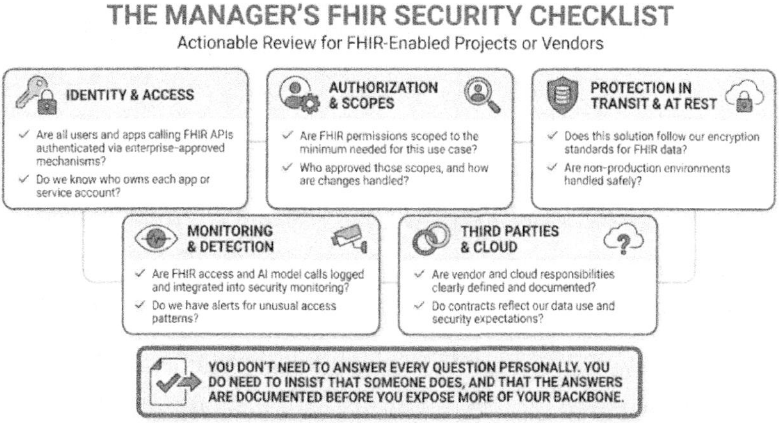

10.12 Chapter 10 Takeaways

- FHIR turns your data into a backbone; without security, that backbone becomes a liability instead of an asset.
- Securing FHIR is about layered controls—identity, authorization, protection, monitoring, and governance—not about a single tool or product.
- Your AI and interoperability ambitions depend on a FHIR environment that leadership, clinicians, and partners can trust; security is how you earn and maintain that trust.

11 The Road Ahead — The Learning Health System

Beyond Projects: Building a Living System

By this point, you've seen FHIR, AI, workflows, governance, security, and pilots from every angle. You have mental models, playbooks, and checklists. The natural question now is: **what are you actually building?** Is this just a collection of better projects, or something more permanent?

The destination is a **learning health system**—not as a buzzword, but as a practical description of how your organization operates. In a learning system, every interaction, every decision, every outcome becomes fuel for improvement. Data flows reliably through FHIR. AI applies intelligence where it adds value. Governance and security keep everything inside safe boundaries. Managers like you steer change based on evidence, not just instinct.

This chapter looks at the road ahead through that lens. You'll see what a learning health system looks like in practice, what capabilities you're quietly assembling, and how to guide your organization over the next few years without pretending you can transform everything overnight.

What a Learning Health System Actually Looks Like

A learning health system is not a single platform you buy. It's a **pattern of behavior** across people, process, and technology. You know you're moving in that direction when:

- Data from care, operations, and finance flows into a shared backbone (FHIR + data platform) as a matter of course, not as a special project.
- AI-powered insights are embedded in everyday tools—huddles, queues, views—not sitting in standalone dashboards nobody checks.
- Outcomes are measured and fed back into design decisions; pilots aren't just celebrated, they are interrogated.
- Governance and security are part of how you do AI and interoperability, not after-the-fact gatekeepers.
- Staff expect workflows to evolve based on what the organization learns, not stay frozen for years.

On the ground, that might look like:

- A medicine unit that routinely uses FHIR-based risk scores, updated daily, to shape discharge and follow-up planning—and adjusts thresholds based on monthly reviews.
- A revenue team that uses AI-assisted denial prediction as a normal part of their process—tuning models and checklists as payer behaviors change.

- A care management program that continuously refines who gets outreach and how, based on what actually improves outcomes for different subgroups.

The key is not perfection; it's **trajectory**. Are you better at learning from yourself this year than last?

The Learning Health System Circle

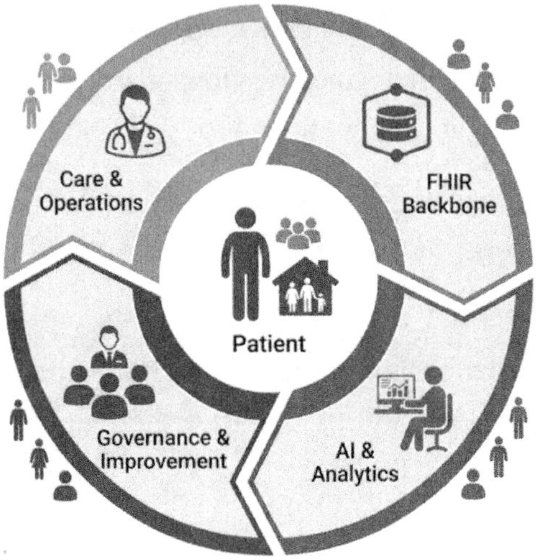

Capabilities You're Quietly Assembling

As you've worked through FHIR and AI, you've been building a set of **organizational capabilities**, whether you named them that way or not:

- **Data as an operational asset**
  You're moving from data as a byproduct of systems to data as something you design and govern (FHIR resources, profiles, quality targets).

- **Reusable AI patterns**
  Prediction, prioritization, classification, and generation are becoming patterns you can reuse across workflows instead of one-off experiments.

- **A governed backbone**
  Security, consent, access control, and responsible AI are becoming standard practice for your FHIR and AI efforts.

- **Pilot-to-production discipline**
  You're learning to distinguish experimentation from scaling, and to harden what works rather than scattering half-finished pilots.

- **Cross-functional ways of working**
  Clinicians, operations, IT, data, security, and compliance are gradually learning to work together on FHIR + AI initiatives rather than in parallel tracks.

These capabilities are cumulative. Each new initiative can be easier and safer than the last if you **reuse what you've already built**—FHIR resources, data pipelines, governance patterns, and change management approaches.

Maturity Stages: Knowing Where You Are

Not every organization will move at the same pace. It helps to think in rough maturity stages so you can be honest about where you are and what comes next.

A simple four-stage view:

- **Stage 1: Ad hoc**
  FHIR and AI appear in isolated projects. Data is extracted on demand. Governance is informal. Success depends on individual champions.

- **Stage 2: Emerging backbone**
  A core set of FHIR resources is implemented and reused. Some AI use cases run on that backbone. Basic governance and security patterns exist, but are still maturing.

- **Stage 3: Defined operating model**
  FHIR + AI are part of a clear operating model. Most new initiatives plug into the backbone. Pilot-to-production patterns are recognized. Governance is predictable.

- **Stage 4: Embedded learning system**
  Continuous measurement, monitoring, and improvement are normal. FHIR and AI capabilities are broadly trusted. Responsible AI and security are cultural expectations, not just policies.

Your job is not to jump from Stage 1 to Stage 4 in a year. It's to:

- Be explicit about where you are.

- Choose a few capabilities to strengthen each year.

- Communicate progress in terms that leaders and staff can feel, not just see on architecture diagrams.

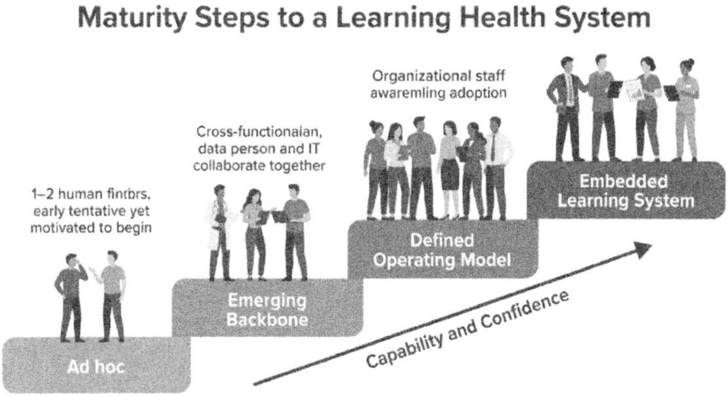

The Next 24–36 Months: A Realistic Agenda

Over the next two to three years, most serious managers can realistically drive progress along a few key dimensions. You don't need to do everything at once. You do need to **pick your moves**.

You might focus on:

- **Deepening the backbone**
  Expand and harden FHIR coverage for 2–4 high-value domains (e.g., inpatient episodes, chronic disease management, revenue-critical workflows). Improve data quality and governance for those resources.
- **Consolidating AI use cases**
  Move from many small AI experiments to a smaller number of flagship workflows you operate well: one or two in clinical care, one or two in operations/revenue.

- **Strengthening guardrails**
  Make your responsible AI and FHIR security practices more routine: a simple risk-tiering approach, standard model documentation, basic fairness checks for higher-risk use cases, and consistent FHIR access controls.

- **Building people's confidence**
  Invest in FHIR + AI literacy for managers and frontline leaders: short sessions, real examples, and clear stories about how these tools support—not replace—them.

- **Aligning vendors to your model**
  Use upcoming procurement and renewal cycles to push vendors toward your FHIR + AI architecture and governance expectations.

Each year, you can ask: "What are the three FHIR + AI capabilities we'll be proud to own at the end of this year that we don't have today?" Then focus ruthlessly.

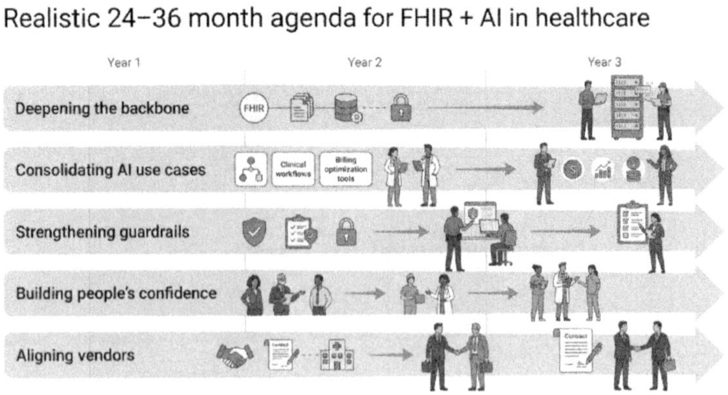

Leading Culture: Making Learning the Default

A learning health system ultimately rests on culture. Technology and governance create the conditions, but people decide whether to use them.

The cultural shift you're aiming for looks like this:

- Teams expect to see **data about their own performance**, and they use it to iterate, not to defend themselves.
- Clinicians and staff feel **safe raising concerns** about AI behavior, data quality, or workflow fit, and see those concerns addressed.
- Managers routinely ask, "What did we learn from this?" when a project goes well or badly.
- Executives see FHIR + AI as part of how the organization learns and adapts, not as special projects for the innovation team.

You can model that by:

- Sharing both successes and adjustments from your FHIR + AI initiatives—what worked, what you changed, what you stopped.
- Treating near-misses as valuable warnings, not as reasons to clamp down on new ideas.

- Recognizing and promoting people who help bridge clinical, operational, and technical worlds.

Culture doesn't change because of a slogan. It changes because people repeatedly see that learning is expected, supported, and put into practice.

Humans Around the Learning Table

Your Role Going Forward

Your role as a "serious manager" does not end with this book. If anything, it becomes clearer.

In the years ahead, you will:

- Translate mission and strategy into **concrete FHIR + AI use cases** that matter.
- Protect and strengthen your **backbone**—data, governance, security—when shortcuts are tempting.

- Demand evidence of **impact and responsibility**, not just demos and promises.
- Connect people across silos so FHIR + AI work is interdisciplinary by default.
- Keep the focus on **workflow-level impact**, where clinicians, staff, and patients actually feel the difference.

You don't need to have all the answers. You do need to ask better questions, set clearer expectations, and keep your organization moving from ad hoc experimentation toward a learning health system it can trust and afford to maintain.

If you keep that posture—mission-aligned, operationally grounded, and responsible by design—you will be ready for whatever comes next: new regulations, new AI capabilities, new demands from patients and payers. You won't just be reacting to the AI and interoperability landscape; you'll be shaping how your organization lives in it.

12 The Technical Hierarchy of FHIR: From Building Blocks to Working Systems

Opening Scenario: "Why Does FHIR Feel So Complicated?"

On a Thursday afternoon, your integration lead walks into your office with a diagram that looks like a circuit board. Boxes labeled "Resources," "Profiles," "Value Sets," "Code Systems," "Extensions," "Operations," and "IGs" are connected by a web of arrows. The team insists, "This is how FHIR really works." You glance at it and think, "I thought FHIR was supposed to simplify interoperability." *Why does the picture look more complex than the systems we already have?*

This chapter exists to decode that picture. FHIR *does* simplify interoperability—but it does so by providing a **technical hierarchy** of components that each play a clear role. If you only see isolated parts ("we send FHIR resources"), everything feels ad hoc. Once you see the hierarchy—how resources, profiles, terminology, operations, bundles, and implementation guides stack—you gain operational clarity. You can tell where a problem lives, what can be standardized once, and what needs local decision-making.

We'll walk the FHIR hierarchy from the bottom up, in practical layers: the **base specification**, **core resources**,

terminology, **constraints and profiles**, **operations and interactions**, **bundles and workflows**, **implementation guides**, and **real-world endpoints and ecosystems**. By the end, a messy diagram will turn into a mental map you can use in every FHIR conversation.

FHIR as a Layered Building

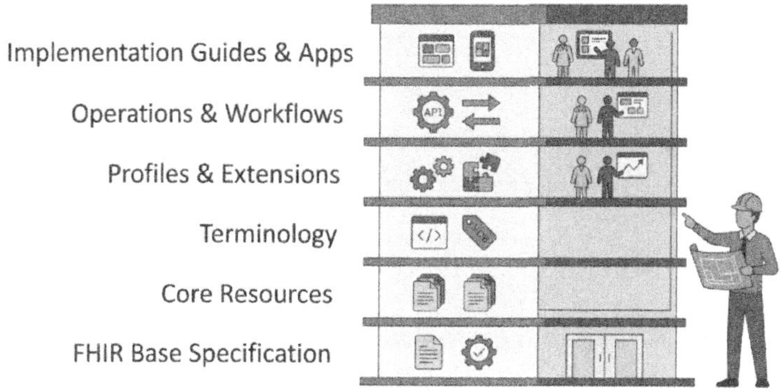

Why a Technical Hierarchy Matters for Managers

You are accountable for outcomes, not for writing FHIR JSON. So why spend time on hierarchy? Because it tells you **where** decisions really live and **what type of work** you're approving:

- When a vendor says "we support FHIR," you can ask: *At which layers? Core resources only, or profiles and implementation guides as well?*
- When teams argue about "FHIR problems," you can localize: *Is this a resource design issue, a terminology issue, a profile issue, or a workflow issue?*

- When you plan a roadmap, you can sequence: *Which layers do we invest in first for our use cases?*

Think of the hierarchy as a management blueprint:

- Lower layers (base spec, resources, terminology) are **foundational standards** you mostly adopt.
- Middle layers (profiles, extensions, operations) are **where your organization shapes FHIR** to your reality.
- Upper layers (implementation guides, APIs, applications) are **where value shows up in workflows** and vendor products.

If you treat everything as "FHIR work" at once, you end up with blurred responsibilities and slow progress. If you align teams and vendors to layers, you get focus and reusable assets.

The Base Specification: FHIR's "Constitution."

At the bottom of the hierarchy is the **FHIR Base Specification**—the document set that defines how FHIR behaves at the most general level. It covers:

- **Data types**: strings, dates, codes, identifiers, references, addresses, quantities, etc.
- **Common elements**: like id, meta, text, and extension that almost every resource can have.

- **Exchange formats**: JSON, XML, RDF—how FHIR is represented on the wire.
- **RESTful API rules**: how resources are created, read, updated, searched, and deleted.
- **Versioning rules**: how changes are handled, what "R4" vs "R5" means, how clients and servers signal capabilities.

You can think of the base specification as the **grammar** of the FHIR language. Every other layer uses this grammar:

- Resources are nouns built from base data types.
- Profiles are detailed "sentences" that must still follow that grammar.
- Operations and interactions use the HTTP verbs and exchange patterns defined here.

As a manager, you don't need to read the spec cover to cover, but you should know:

- Which **FHIR version** your systems and vendors use (e.g., R4 vs R5).
- Whether teams are aligning to **standard exchange rules** or inventing custom APIs "on top of FHIR data."
- Who owns your organization's position on **version upgrades** and compatibility?

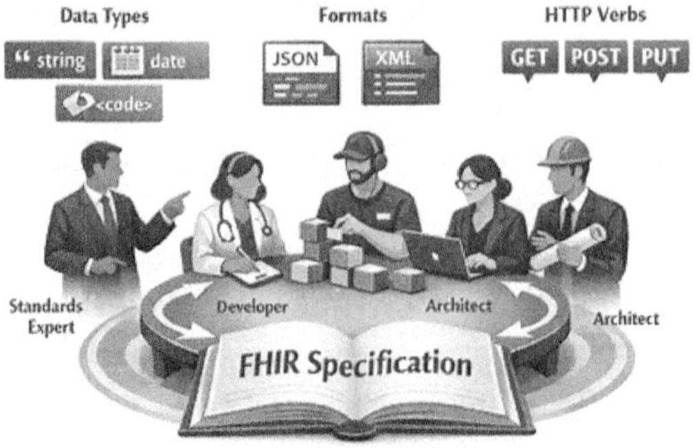

FHIR Base Spec as Grammar

Core Resources: The "Nouns" of Healthcare Data

Above the base spec sit **resources**—the core building blocks of FHIR. A resource is a small, reusable package of healthcare information with a defined structure and meaning. Examples you will see in almost every project:

- **Clinical**: Patient, Encounter, Condition, Observation, MedicationRequest, Procedure, AllergyIntolerance.
- **Administrative**: Organization, Practitioner, Location, Coverage, Claim.
- **Workflow**: Task, ServiceRequest, Appointment.
- **Infrastructure/meta**: Bundle, OperationOutcome, CapabilityStatement.

FHIR's design philosophy is many small, composable resources rather than **a few giant messages**. Rather than send "the chart," you send a set of related resources:

- A patient's latest vitals: Patient + multiple Observation resources.
- A referral: Patient, ServiceRequest, Condition, Practitioner, Organization.
- A claim: Claim + references to Patient, Encounter, Coverage, etc.

Each resource:

- Has a stable URL name (e.g., /Patient, /Observation).
- Defines required and optional fields.
- Uses base data types and common patterns from the spec.

Manager implications:

- When you hear "we support FHIR," ask: *Which resources, and with what level of completeness?*
- Resource coverage defines what **questions you can answer** and what workflows you can support.
- Most "we can't do that yet" stories boil down to: **the right resource or field isn't populated or exposed**.

3.5 Terminology: Code Systems, Value Sets, and Concept Maps

Resources define structure; **terminology** defines meaning. FHIR's terminology layer sits beside resources and powers data quality, analytics, and interoperability. It includes three main components:

- **Code systems** – The vocabularies themselves (e.g., SNOMED CT, LOINC, ICD-10, RxNorm, internal code lists).
- **Value sets** – Named *subsets* of one or more code systems, designed for specific use (e.g., "active smoking status codes," "diabetes diagnosis codes").
- **Concept maps** – Explicit mappings between codes in different systems (e.g., mapping a proprietary lab code to LOINC).

In practice:

- A FHIR Observation might carry a LOINC code for the lab test and a SNOMED CT or SNOMED-like code for interpretation.
- A Condition uses standard diagnosis codes so that AI models and analytics can scale across sites.
- A value set defines which codes count as "hypertension" or "depression" for a measure or algorithm.

Terminology is where **semantic consistency** lives. Two organizations can both expose Observation resources, but if they code the same test differently, AI and reporting will disagree.

Manager implications:

- When planning AI or quality programs, ask: *Which code systems and value sets are canonical for this use case? Who owns them?*
- Terminology governance is not optional—the same resource plus different codes = different outcomes.
- FHIR gives you a *place* to put standard codes; it does not solve your coding discipline for you.

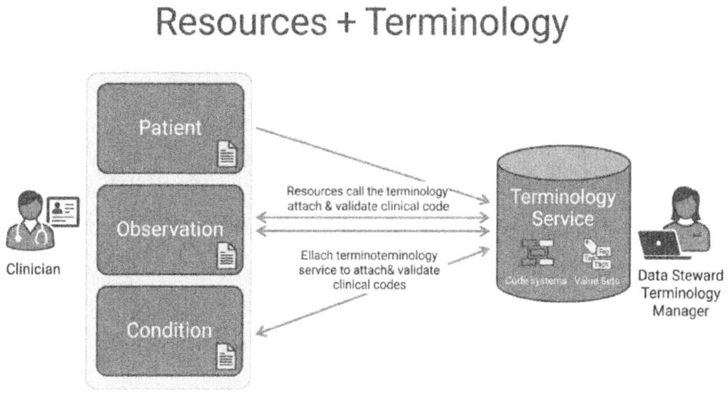

Profiles: Turning Generic FHIR into "FHIR for Us."

Out of the box, FHIR resources are **generic**. That's intentional—they need to work from rural clinics to national programs. In real implementations, you almost always need **profiles**: formal constraints that adapt a generic resource to a specific program, jurisdiction, or workflow.

A **FHIR profile** can:

- Mark elements as **required** or **optional** for your use case.
- Restrict data types or cardinality (e.g., “exactly one identifier of type MRN”).
- Bind fields to specific **value sets** (e.g., “these are the only allowed codes for Condition. code in this program”).
- Define specific **extensions** that add new fields (more on extensions next).

Examples:

- A national immunization program profile on Immunization that requires the batch number and route.
- A payer profile on Coverage that requires plan ID and network details.
- A research profile on Observation that requires specimen information.

Profiles are where “FHIR in theory” becomes **FHIR in your contract, policy, or workflow**.

Manager implications:

- Ask vendors and internal teams: *Which profiles are you using or supporting? Are they standard (e.g., US Core, national IGs) or home-grown?*
- Profiles are **reusable assets**. Investing once in good profiles for core workflows pays off across projects.

- Misalignment often happens when each project creates its own profiles without coordination.

Extensions: Adding What FHIR Didn't Predict

No standard can anticipate every local need. FHIR's answer is **extensions**: a safe, structured way to extend resources without breaking compatibility.

An extension:

- Lives at a specific path in a resource (e.g., an extension on Patient or Observation.value).
- Has its own definition (a "mini-profile" for that additional data).
- Is identified by a URL so others can recognize and reuse it.

Used well:

- Extensions capture *local* or *emerging* requirements (e.g., local consent flags, internal triage categories, AI-derived scores before a standard exists).
- Over time, widely used extensions can inform future versions of the standard.

Used poorly:

- Extensions replicate fields that already exist in core resources.

- Each project invents its own extension for the same concept, undermining interoperability.

Manager implications:

- Ask: *What proportion of our data is in standard elements vs extensions? Are we extending where we must, or because we haven't done the governance work?*
- Require that extensions be **cataloged and reviewed**, especially if they relate to clinical safety, consent, or payment.
- Treat extension sprawl as an architectural smell—like proliferation of one-off APIs.

Operations, REST Interactions, and Search: How Systems Actually Talk

Resources and profiles define **what** you can exchange; operations and interactions define **how** systems talk. The FHIR base spec includes:

- **RESTful interactions**: standard CRUD and search operations:
  - CREATE (POST), READ (GET), UPDATE (PUT/PATCH), DELETE (DELETE).
  - SEARCH with rich parameters (e.g., find Observation by patient, code, date, etc.).

- **Batch and transaction** interactions: sending multiple resources in a single Bundle.
- **Operations**: special-named operations for more complex behaviors (e.g., $everything, $export, $validate).

This layer is where you see FHIR as an **API standard**, not just a data standard. For example:

- A care management app calls GET /Patient/{id} and GET /Observation?patient={id}&code=....
- A payer system uses $export to bulk extract patient data for analytics.
- A validation service uses $validate to check resources against profiles.

Manager implications:

- When you fund "FHIR integration," clarify: *Are we only doing data mapping, or also exposing/consuming FHIR APIs according to the spec?*
- Search and operations choices drive **performance, privacy, and governance** concerns.
- Some vendors use "FHIR-like JSON" but don't implement standard operations—call that out explicitly.

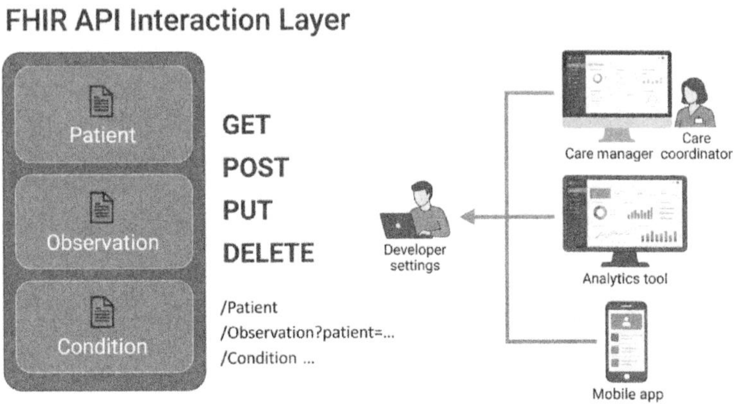

Bundles, Composition, and Workflow Resources: Turning Pieces into Stories

FHIR doesn't just send single resources; it can send **collections** that represent meaningful units of work. Key constructs:

- **Bundle** – A container resource that groups other resources. Types include:
  - collection – e.g., a set of Observation resources.
  - transaction – multiple resources to be processed as a single unit.
  - message – event-like bundles.
  - document – a clinical document with a narrative and supporting resources.

- **Composition** – A resource that defines the structure of a document (e.g., sections in a discharge summary).
- **Workflow resources** – Task, ServiceRequest, Appointment, CarePlan, etc. that drive work, not just record it.

These are where FHIR moves from static data to **clinical and operational stories**:

- A discharge summary as a Bundle (type document) containing Composition, Encounter, Condition, Procedure, MedicationRequest, Observation, etc.
- A care plan represented as a CarePlan with linked Goal and Activity resources.
- A case management queue driven by Task resources assigned to staff.

Manager implications:

- If your teams only exchange single resources, they may be missing opportunities to model **complete workflows and documents**.
- Bundles and workflow resources are where AI and automation can meaningfully plug into clinician-facing flows.
- Ask: *Are we using FHIR only as a data warehouse feed, or as a workflow backbone?*

Implementation Guides: Packaging Rules into Something People Can Use

An **Implementation Guide (IG)** is a package that ties the lower layers together into a usable recipe for a specific domain or program. An IG typically includes:

- A set of **profiles** on key resources.
- **Value sets** and terminology bindings.
- **Examples** of valid payloads.
- **Narrative documentation** on use cases, workflows, and constraints.
- Sometimes **conformance requirements** (e.g., "servers SHALL support these interactions").

Examples you'll see in the field:

- National or regional core guides (e.g., "US Core," country-specific base guides).
- Program-specific guides (e.g., immunization registries, quality reporting, patient access APIs).
- Vendor or ecosystem guides that define how their platform uses FHIR.

Implementation guides are where the FHIR hierarchy becomes **contractual**: they say, "This is how we implement FHIR for this purpose."

Manager implications:

- For any serious FHIR-backed project, ask: *Which IGs are we using or producing? Are they recognized externally?*
- IGs are your best lever for **repeatability**—use them to avoid one-off integrations.
- Supporting the right national or industry IGs makes your systems plug into broader data-sharing and AI ecosystems.

Implementation Guide as Recipe Book

FHIR Servers, Clients, and Ecosystems: How the Pieces Show Up in Real Life

All the layers above become real inside **systems and ecosystems**:

- **FHIR servers** – Components (often part of an EHR or data platform) that store resources and expose them via the FHIR API.
- **FHIR clients** – Apps, services, AI models, and integration tools that read/write FHIR.
- **Intermediaries** – Gateways, adapters, and HIEs that translate between FHIR and non-FHIR systems or mediate access.

In a typical environment:

- Your EHR acts as a FHIR server for core clinical data.
- A population health platform acts as a client—reading from the EHR, enriching data, writing back risk assessments.
- Payers and external partners connect through FHIR APIs to exchange coverage, claims, and prior authorization data.
- AI services consume FHIR resources from one or more servers, then publish derived insights back as new FHIR resources.

Manager implications:

- Ask for an **ecosystem map**: which systems are servers, which are clients, what IGs and profiles they support.
- Clarify **ownership**: who runs the central FHIR backbone vs edge clients.

- When problems arise, use the hierarchy to localize: is this a server conformance issue, a client misuse issue, a profile mismatch, or a terminology gap?

Technical Hierarchy as a Governance Tool

Understanding the FHIR hierarchy also gives you a **governance handle**:

- At the **base spec/core resource** level, you mostly adopt global standards.
- At the **terminology and profiles** level, you make organization-wide decisions that should be centrally coordinated.
- At the **implementation guide and ecosystem** level, you manage projects, vendors, and contracts.

You can structure responsibilities accordingly:

- A small, skilled **FHIR governance group** owns profiles, extensions, terminology, and IG decisions.
- Product and project teams consume those assets, rather than redefining them.
- Vendor selection and RFPs require conformance at specific layers (e.g., "must support US Core R4 profiles for Patient, Practitioner, Encounter…").

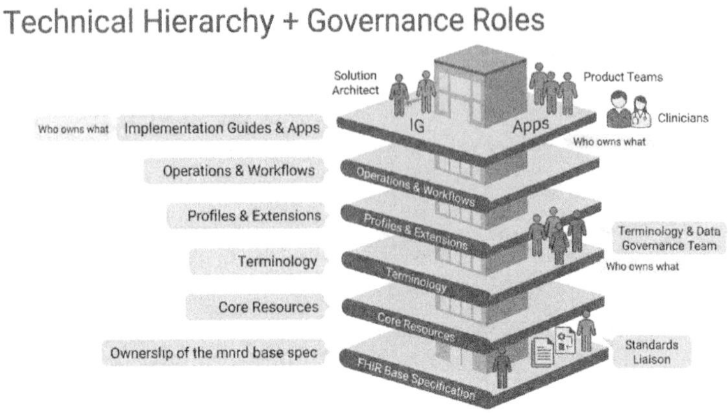

The Manager's Checklist: Navigating the FHIR Hierarchy

When you review a FHIR proposal or integration, use the hierarchy as a structured checklist:

- **Base & Version**
  - Which FHIR version are we standardizing on for this initiative (R4, R5)?
  - Are all participating systems aligned, or is there a version translation plan?
- **Core Resources**
  - Which resources are in scope (Patient, Encounter, Observation, etc.)?
  - Are we clear on which systems are the source of truth for each resource type?
- **Terminology**

- Which code systems and value sets are canonical for this use case?
  - Who owns their governance and updates?
- **Profiles & Extensions**
  - Are we using established profiles (national/industry/core) where possible?
  - Which extensions are truly necessary, and are they documented and governed?
- **Operations & Workflows**
  - Which standard operations and searches will we support or depend on?
  - Are we using bundles and workflow resources to represent real-world processes, not just single records?
- **Implementation Guides & Ecosystem**
  - Which IGs do we conform to, and what is our role (publisher vs consumer)?
  - How do our FHIR servers, clients, and adapters fit into the broader ecosystem map?

If teams can answer these questions in a page or two, you have **operational clarity** over the technical hierarchy. If not, you're likely funding one-off plumbing that will be hard to reuse or scale—especially when AI enters the picture.

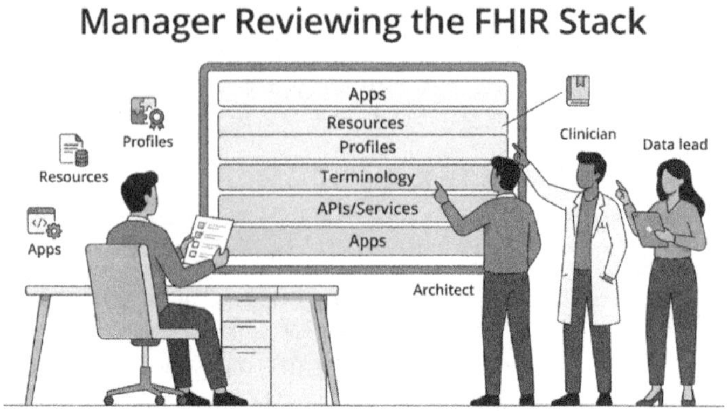

Chapter Takeaways

- FHIR is not one thing; it is a **technical hierarchy**: base specification, resources, terminology, profiles/extensions, operations/workflows, and implementation guides.
- Each layer has a distinct purpose and ownership pattern—recognizing that prevents confusion and duplication.
- Resources without shared terminology, profiles, and IGs are like a language without grammar or dictionaries: technically valid, operationally fragile.
- As a manager, your leverage comes from asking **layered questions** and insisting that teams and vendors anchor their designs to the right parts of the FHIR hierarchy.

13 Governing AI-on-FHIR Workflows End to End

Opening Scenario: "Who Approved What This AI Just Did?"

It's 8:15 a.m. A clinician flags something odd in your new AI-on-FHIR workflow. A patient's record shows a high-risk readmission score, a new outreach Task, and a message queued for the call center—all created automatically overnight. The clinician agrees the patient needs attention, but asks you a simple, uncomfortable question: *"Who approved this? The model? The system? Me?"*

Your logs show:

- FHIR resources created by an AI service account.
- Data pulled from multiple systems via FHIR.
- No obvious record of which model version or policy was in force.

The technical integration "works." The governance story is fuzzy.

This chapter closes that gap. You'll see how to govern AI-on-FHIR workflows **end to end**—from identity and consent through provenance, logging, and human-in-the-loop—so you can answer, in plain language, "who did what, with whose authority, based on which data."

Why AI-on-FHIR Governance Is Not Just "More Security"

When you connect AI and FHIR, governance gets more complex for three reasons:

- **More actors** – Humans, AI services, orchestrators, and external vendors can all read and write FHIR resources.
- **More automation** – Systems can create tasks, flags, and messages without a person clicking a button.
- **More consequences** – Decisions informed or initiated by AI can affect care, payment, and reputation faster and at scale.

Traditional controls—RBAC, audit logs, and change management—are necessary but incomplete. You need a **workflow-level view** that explicitly governs:

- Which AI systems are allowed to act on which FHIR data?
- How patient consent and policy affect those actions.
- How human oversight is structured and documented.
- How do you trace and correct AI-driven actions when something goes wrong?

Think of this chapter as your **governance overlay** on the technical hierarchy and AI patterns from earlier chapters.

Governance Lens Over AI-on-FHIR

Identity and Roles: Who Is Allowed to Act on FHIR?

Governance starts with **identity**. In AI-on-FHIR workflows, you typically have four identity types:

- **Patients and members** – Subjects of the data and, increasingly, active participants (portals, apps).
- **Clinicians and staff** – Direct users of AI-augmented workflows.
- **AI services and orchestrators** – Non-human actors that read and write FHIR resources.
- **Vendors and external partners** – Third-party systems accessing your FHIR endpoints.

For each identity type, you need clarity on:

- What resources they can see (e.g., Observation, Condition, RiskAssessment).
- What actions they can take (read-only vs create/update).
- Under what conditions (e.g., only in certain care settings, tenants, or programs).

In practice, this means:

- AI services and orchestrators are treated as **first-class clients** with their own credentials, scopes, and policies—not "trusted back-end jobs" with unlimited access.
- Clinician actions taken *because of* AI outputs are still **clinician actions**, but the AI's role is recorded.
- Vendor AI systems are given **bounded, revocable access** to well-defined FHIR APIs, not raw database feeds.

Your governance question: *Can we list, in one place, all AI-related identities, what FHIR data they can access, and what they are allowed to change?*

Consent, Purpose, and Legal Basis: When Is AI Allowed to Use FHIR Data?

FHIR makes it easier to move and structure data. That doesn't automatically make it acceptable to use that data for any AI purpose. Governance must connect:

- **Consent and policy** (what patients and regulations allow).
- **Purpose** (clinical care, operations, research, quality, payment).
- **Actual AI behavior** (what models and workflows do with the data).

Key design decisions:

- How you represent **consent and restrictions** in FHIR (e.g., Consent resources, meta. security tags).
- How AI services **check consent** before accessing or using data.
- How you handle **secondary use** (e.g., using care data to train models for future use).

For operational AI-on-FHIR workflows, you should be able to answer:

- Is this AI acting under the same legal basis as the underlying care or payment workflow?
- If not, what additional consent or approvals are in place?
- How do we prevent AI services from using FHIR data beyond their approved purposes?

"Consent and Purpose in AI-on-FHIR"

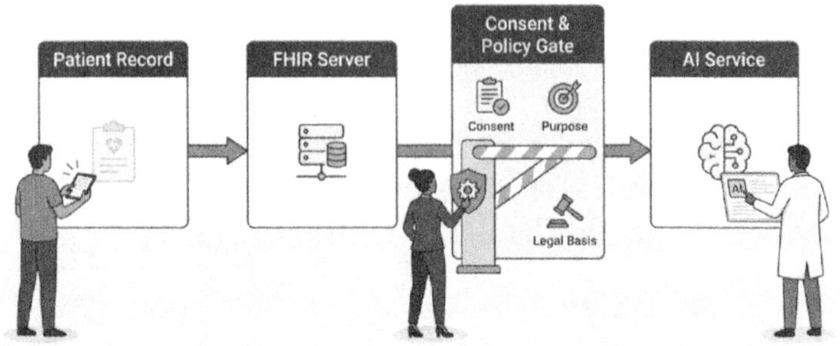

Provenance and Traceability: Being Able to Reconstruct What Happened

When AI writes into FHIR—risk scores, flags, tasks, drafts—you need a way to say **who created what, based on which inputs, with which model.** This is where **provenance** and related metadata matter.

Practical elements:

- Tag AI-derived resources with:
    - The **actor** (AI client identity).
    - The **model or rule set** (version identifier).
    - The **timestamp** and context (e.g., job run, request ID).
- Use FHIR Provenance and/or AuditEvent resources to link outputs to inputs:

- Which Condition, Observation, Encounter resources were used?
- Whether a human reviewed or modified the output before it was acted on.

Benefits:

- When someone asks, "Why was this patient flagged?", you can show both the underlying clinical data and the AI trace.
- When a model is updated, you can see which outputs were created under which version.
- When investigating incidents, you can distinguish **model failure** from **input data issues** or **misuse**.

Governance requirement: make provenance **non-negotiable** for AI-on-FHIR writes, especially for high-impact or high-risk workflows.

Human-in-the-Loop on a FHIR Backbone

"Human-in-the-loop" becomes real when it is **visible in FHIR workflows**, not just promised in slide decks. That means:

- AI creates **suggestions**, not final clinical decisions or irreversible actions, by default.
- Suggestions are represented as FHIR resources (Task, CommunicationRequest, DetectedIssue, draft notes)

with a clear **status**: proposed, in-progress, completed, rejected.

- Human actions—accepting, editing, or rejecting AI suggestions—are recorded in FHIR, linked to the original suggestion.

Examples:

- A care coordinator sees an AI-generated task: "Call patient about missed follow-up." They can:
    - Accept and complete it (with notes).
    - Defer it.
    - Reject it as not clinically appropriate.
- A clinician sees a draft note created by an AI documentation assistant, stored as a draft Composition. Their edits and sign-off are preserved.

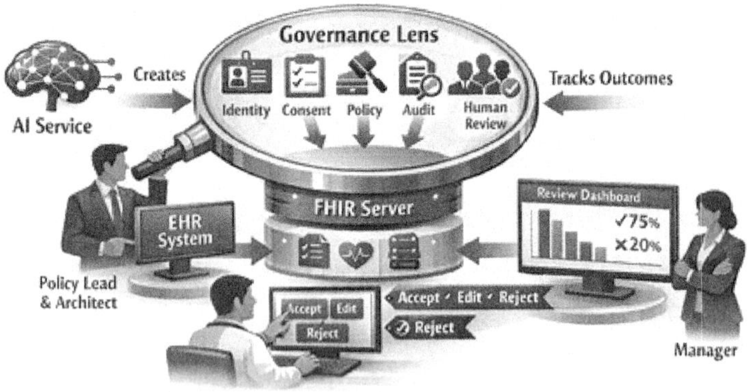

Human-in-the-Loop Workflow in FHIR

Governance questions:

- For each AI-on-FHIR use case, where exactly does human review happen, and how is it recorded?
- Which actions are **never** automated end-to-end without human involvement (e.g., medication changes, major care plan alterations)?
- How are patterns of acceptance/rejection used to improve both models and workflows?

Policy-Backed Guardrails in the FHIR Layer

AI governance policies (e.g., "AI may not access data from certain programs" or "High-risk suggestions must be reviewed by a senior clinician") are only useful if they are **enforced concretely**. FHIR gives you several enforcement points:

- **Authorization rules** – Which AI clients can access which resources, in which scopes.
- **Profile and IG constraints** – What AI is allowed to write (e.g., mandatory fields for risk scores, allowed status values).
- **Terminology and tagging** – Which codes/flags indicate that a resource is AI-derived, under review, or confirmed?

Examples of policy-backed guardrails:

- A rule that AI services cannot create or update MedicationRequest resources directly; they can only propose Task items for clinicians.
- Profiles that require AI-generated RiskAssessment resources to include model version and confidence score.
- Security tags that restrict visibility of certain AI outputs (e.g., behavioral health risk scores) to specific roles.

Your job is to push policy discussions down into the **FHIR configuration layer**:

- "Where does that rule live in our stack?"
- "What FHIR construct enforces it?"
- "How will we verify it stays in place over time?"

5.8 Vendor and Ecosystem Governance: AI-on-FHIR Across Organizational Boundaries

Few organizations run AI-on-FHIR in isolation. Vendors and partners will:

- Expose FHIR APIs you consume.
- Consume your FHIR APIs to run their own AI features.
- Offer AI services that plug into your FHIR backbone.

Governance tasks on your side:

- Require **clear FHIR contracts** in RFPs and agreements:
    - Supported FHIR versions, resources, profiles, and IGs.
    - How their AI services will read and write FHIR data.
- Specify **data handling and AI use**:
    - Which FHIR data can they use for training vs operations?
    - How AI outputs are represented and shared back.
    - What provenance, logging, and explainability do you expect?
- Ensure **revocation and oversight**:
    - Ability to revoke their FHIR access quickly.
    - Access to logs and audit trails for AI usage.

This is not just procurement language—it's how you extend your AI-on-FHIR governance into a **multi-party ecosystem** without losing visibility or control.

Monitoring and Incident Response in AI-on-FHIR Workflows

Monitoring and incident response change when AI is part of your FHIR workflows. You're watching for issues at multiple layers:

- **Data layer** – sudden changes in coding, missing fields, unusual patterns in FHIR resources.
- **Model layer** – performance drift, fairness issues, atypical outputs.
- **Workflow layer** – spikes in AI-generated tasks, unusual acceptance/rejection patterns, delays in human review.

You'll want dashboards that answer questions like:

- How many AI-generated RiskAssessment resources were created this week, by which models, for which programs?
- What percentage of AI-suggested Tasks were accepted vs rejected, by role, unit, or site?
- Are there anomalies in AI access to sensitive resources (e.g., sudden increases in certain Observation queries)?

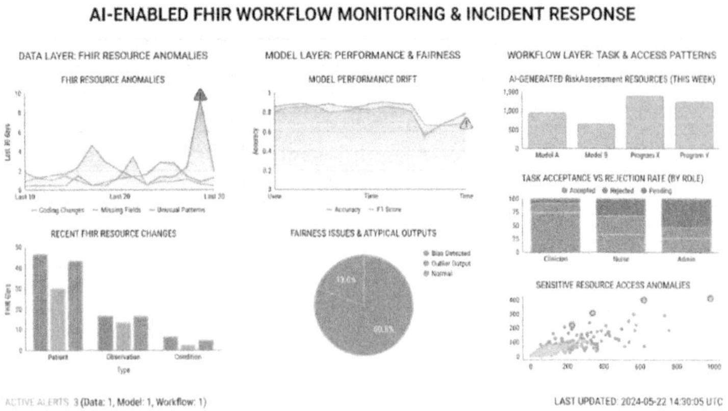

When incidents occur (e.g., an AI model mislabels a cohort, or an orchestrator creates too many tasks):

- Use FHIR provenance and audit data to **reconstruct the chain**.
- Decide whether to **disable AI writes**, remove or correct affected resources, and/or adjust visibility.
- Feed lessons back into both model governance and FHIR configuration (profiles, policies, access).

Manager's AI-on-FHIR Governance Checklist

For each AI-on-FHIR workflow, insist on these answers:

- **Identity & Access**
    - Which human and system identities are involved, and what can each read/write in FHIR?
    - Are AI services and orchestrators treated as first-class clients with scoped access?
- **Consent & Purpose**
    - What legal basis and consents apply?
    - How does the AI layer enforce those constraints in practice?
- **Provenance & Traceability**

    - How do we trace AI outputs back to inputs, models, and human reviewers?
    - Are we using FHIR constructs (e.g., Provenance, AuditEvent) consistently?
- **Human-in-the-Loop**
    - Where is human review required, and how is it captured in FHIR resources and statuses?
    - Which actions can never be fully automated by AI?
- **Policy Enforcement & Vendors**
    - Which policies are enforced in FHIR (profiles, tags, access rules), not just in documentation?
    - How are vendors bound to these same expectations?

If your teams cannot answer these questions clearly, you are likely running AI-on-FHIR workflows that **are technically sound but under-governed**. That is where reputational, clinical, and regulatory surprises live.

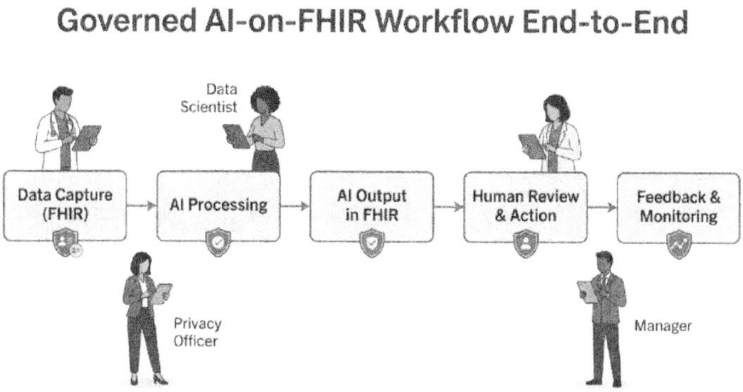

Chapter Takeaways

- AI-on-FHIR governance is about **workflows**, not just APIs and models. You must see the whole chain from data to decision to outcome.
- Identities, consent, provenance, and human-in-the-loop need **concrete representations** in FHIR and surrounding systems, not just policies on paper.
- FHIR can encode and enforce many of your AI guardrails, but only if you deliberately push policies into profiles, tags, access rules, and workflows.
- Vendors and partners must be brought under the same governance umbrella, with FHIR contracts and AI usage expectations that are explicit and enforceable.

14 Scaling AI-on-FHIR Across the Enterprise

Opening Scenario: From One Great Pilot to Ten Inconsistent Projects

Your first AI-on-FHIR project is a success. A readmission risk model writes RiskAssessment resources into your FHIR server. Clinicians see prioritized lists. Feedback is positive. Leadership asks the inevitable question: *"Great—can we do the same thing for sepsis, care gaps, denials, and staffing?"*

Three months later, you have:

- A vendor-driven sepsis score using its own FHIR flavor.
- A care-gap analytics team using bulk FHIR export but not writing back.
- A revenue cycle pilot taking data from a separate warehouse with "FHIR-like" structures.

Each team insists its approach is "just a quick variant." You see a creeping pattern: **every new AI-on-FHIR use case is becoming its own mini-architecture**. This chapter gives you a way out of that trap. We'll focus on intentionally scaling AI-on-FHIR, using reusable patterns, shared infrastructure, and a simple operating model that keeps teams moving fast without fragmenting your FHIR backbone.

Why Scaling AI-on-FHIR Is Harder Than Adding More Models

Scaling is not just "more of the same." When you go from one AI-on-FHIR workflow to many, three challenges appear:

- **Variation** – Different specialties, programs, and vendors want different data, profiles, and workflows.
- **Velocity** – New ideas arrive faster than your central teams can handle if everything is bespoke.
- **Visibility** – It becomes harder to see where AI touches FHIR, what models exist, and who owns them.

If you treat each project as a one-off, you'll get:

- Conflicting profiles and extensions.
- Multiple "truths" about patients and risk across systems.
- Governance that relies on heroics and memory instead of structure.

Your goal is a **repeatable AI-on-FHIR operating model**: shared assets for common needs, clear rules for variation, and lightweight controls that keep AI and FHIR aligned as the portfolio grows.

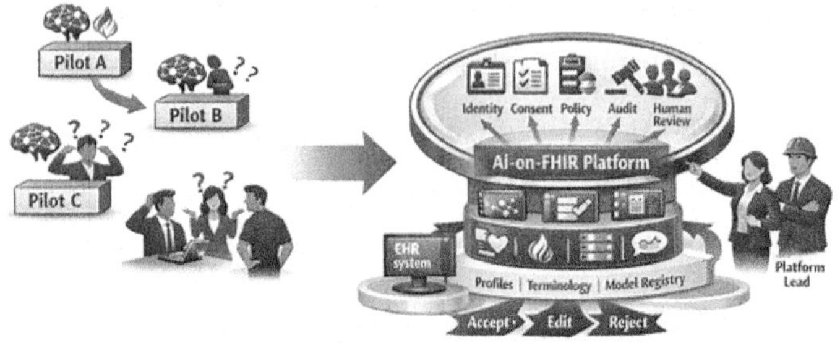

From One-Off Pilots to a Shared Platform

The AI-on-FHIR Platform Mindset

To scale, you need to think beyond "projects" and build an **AI-on-FHIR platform** mindset. That doesn't necessarily mean a single technology product; it means a **set of shared capabilities** and expectations:

- **Shared FHIR backbone** – Agreed versions, core resources, profiles, and terminology for the majority of use cases.
- **Standard AI data access patterns** – How models read from and write to FHIR, with consistent security and provenance.
- **Model registry and catalog** – A simple inventory of models that touch FHIR: what they do, who owns them, which resources they use.
- **Monitoring and observability** – Common logging and dashboards for AI-on-FHIR activity across the estate.

The platform mindset answers the question: *"If a new AI idea comes in tomorrow, what's the standard way it plugs into FHIR?"* If you can't answer that, scaling will be painful.

Reusable Building Blocks for AI-on-FHIR

Instead of treating every AI feature as a blank slate, invest in **reusable building blocks** that teams can assemble. Think in four categories:

1. **Data Building Blocks**
    - Canonical FHIR profiles for high-value domains (e.g., chronic conditions, labs, medications, encounters).
    - Standard value sets for risk factors, care gaps, and outcome labels.
    - Shared extraction/query patterns (e.g., "cohort definition for heart failure patients").
2. **AI Integration Patterns**
    - Templates for read-only analytics, embedded insights, and orchestrated workflows.
    - Sample code or configuration showing how to read/write FHIR with provenance and security in place.
3. **Governance Artifacts**

- Standard AI-on-FHIR design template: resources used, profiles, consent model, human review, outputs.
- Standard risk assessment questions for AI-on-FHIR use cases.

4. **Operational Services**
    - A small set of shared services: FHIR gateway, terminology service, model registry, monitoring pipeline.
    - Reusable "AI writer" component that converts model outputs into FHIR resources consistently.

Your message to teams: **"Don't invent new plumbing. Use the standard pieces and focus your creativity on the clinical or business logic."**

A Simple AI-on-FHIR Operating Model

Scaling requires a **clear operating model**: who does what, at which layer, for every new AI-on-FHIR initiative. A pragmatic structure:

- **Platform / FHIR Team**
  - Owns FHIR infrastructure, core resources, profiles, terminology, and security.
  - Offers patterns and APIs for AI access and write-back.
- **AI / Data Science Team**
  - Designs, trains, and maintains models.
  - Works with the platform team to map inputs/outputs to FHIR.
  - Registers models and their FHIR dependencies.
- **Product / Workflow Owners**
  - Own the use case and outcomes.
  - Decide how AI outputs appear in workflows and where human review sits.
  - Ensure alignment with clinical, operational, and regulatory goals.
- **Governance Group (AI + Data + Clinical + Privacy)**
  - Approves higher-risk AI-on-FHIR designs.

- Sets standards for consent, provenance, and human-in-the-loop.
- Reviews monitoring and incidents.

You don't need a large bureaucracy; you need **habitual collaboration** among these roles. Every AI-on-FHIR project should have named counterparts in each group.

6.6 Standardizing AI-on-FHIR Design Reviews

To avoid long, bespoke reviews for each project, standardize the **questions and artifacts**:

For every AI-on-FHIR proposal, expect a brief design pack that answers:

- **Use Case & Pattern**
  - What problem are we solving, and which AI-on-FHIR pattern (analytics, embedded, orchestration) are we using?
- **FHIR Interface**
  - Which resources and profiles are inputs?
  - Which resources are outputs, with what profiles?
- **Governance Hooks**
  - How are consent and purpose enforced?
  - How is human review structured and recorded?

- **Operations & Risk**
    - How will we monitor the model and its FHIR interactions?
    - What's the rollback/kill switch if something goes wrong?

Review sessions can then focus on **exceptions** instead of re-litigating the basics each time.

Managing Variation Without Losing Control

You can't standardize everything. Different specialties, programs, and jurisdictions will need tailored profiles, workflows, and AI behaviors. The trick is to **manage variation** intentionally:

Define three "rings" of standardization:

1. **Core (must be shared)**

- FHIR version and basic resource definitions.
- Common patient and encounter profiles.
- Core terminology for diagnoses, labs, and medications where possible.

2. **Program-Level (shared within a domain)**
   - Profiles and value sets for specific domains (e.g., oncology, cardiology, behavioral health).
   - AI-on-FHIR patterns tuned for that domain (typical inputs and outputs).
3. **Local (allowed variation)**
   - Extensions for genuinely local needs.
   - Workflow tweaks (e.g., status transitions, notification rules).

Governance principle:

- **Core** changes require enterprise-level review.
- **Program-level** changes require domain-level review.
- **Local** changes require documentation and basic checks, but not full committee cycles—unless risk is high.

This lets you say "yes" to local innovation, while protecting the structural elements your AI-on-FHIR ecosystem depends on.

Coordinating Releases: FHIR Changes and Model Changes

At scale, you'll constantly evolve both FHIR and AI. You need a **joint release discipline** so one doesn't break the other.

Key practices:

- Maintain a **dependency map**: which models depend on which FHIR resources, profiles, and value sets.
- For any FHIR schema or terminology change, identify affected models and decide:
    - Do we retrain or at least revalidate?
    - Do we adapt mappings or features?
- For any major model update, assess FHIR impact:
    - Are we changing resource types, coding, or output profiles?
    - Do downstream systems need updates to display or use outputs correctly?

Create a simple **change-impact playbook**:

- Small FHIR change (e.g., new optional field) → test only.
- Medium FHIR change (e.g., tightened profile) → test + model validation.
- Large FHIR change (e.g., resource coverage) → possible retraining and workflow review.

This avoids two common failure modes:

- FHIR team shipping changes that silently degrade model performance.
- AI team shipping changes that break FHIR consumers or governance assumptions.

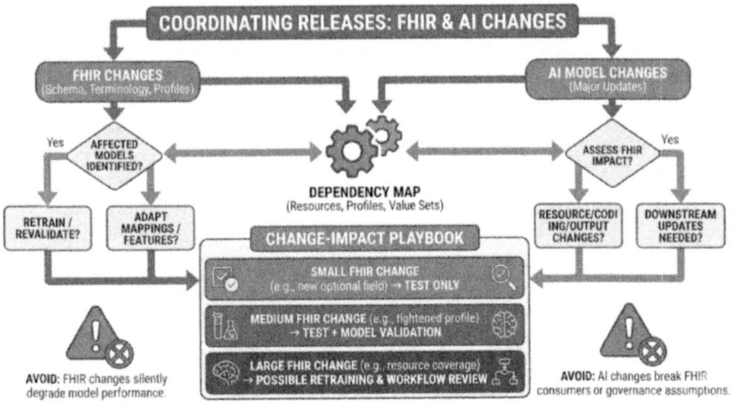

Portfolio View: Seeing All AI-on-FHIR Workflows at Once

Once you have more than a handful of initiatives, you need a **portfolio view** of AI-on-FHIR:

At minimum, track:

- Each model that touches FHIR: purpose, owner, risk tier, pattern (analytics/embedded/orchestrator).
- Input resources/profiles and output resources/profiles.
- Environments in use (dev/test/prod) and rollout status.

- Key governance attributes: consent model, human review requirements, and monitoring status.

Use this view to:

- Identify duplication: multiple teams building similar models on similar FHIR data.
- Spot inconsistent patterns: some models write back to FHIR with provenance, others do not.
- Plan upgrades: which initiatives must be sequenced for a FHIR or platform change?

You don't need an elaborate tool; a well-maintained registry and a simple dashboard are enough to start. The point is visibility, not perfection.

Measured Scaling: KPIs for AI-on-FHIR Maturity

To know whether your scaling approach is working, define a few **AI-on-FHIR KPIs**:

- **Coverage**
  - Percentage of AI initiatives that use the shared FHIR backbone rather than one-off feeds.
  - Percentage of AI outputs represented as FHIR resources (for eligible use cases).
- **Consistency**
  - Number of AI-on-FHIR workflows using standard patterns and profiles.

    - Number of deviations/exceptions and whether they are shrinking over time.
- **Safety & Quality**
    - Frequency and severity of AI-on-FHIR related incidents.
    - Rate of AI suggestion acceptance vs rejection across key workflows.
- **Speed**
    - Average time from AI-on-FHIR idea to approved design using the standard patterns.
    - Impact of platform investments on cycle time (e.g., "second and third projects went faster").

Share these metrics with leadership to show that investing in **shared AI-on-FHIR capabilities** delivers both control and speed.

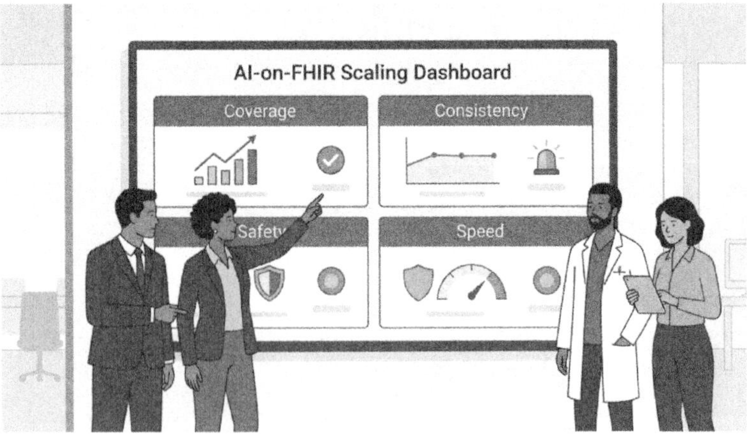

Avoiding Scaling Pitfalls

As you scale, watch for these common pitfalls:

- **Shadow AI-on-FHIR** – Teams quietly connecting models to FHIR without using shared patterns or governance.
- **Profile proliferation** – Every project creates its own profiles and extensions for similar concepts.
- **Vendor fragmentation** – Each vendor solution uses its own FHIR variant and AI integration, creating parallel universes.
- **Governance overload** – Reviews become so heavy that teams route around them or defer AI-on-FHIR entirely.

Countermeasures:

- Make it **easier** to use the standard platform and patterns than to go rogue.
- Provide quick, predictable review for low-risk, pattern-aligned use cases.
- Use contracts and RFPs to align vendors to your FHIR and AI expectations.
- Periodically prune and consolidate profiles, patterns, and tools that are not pulling their weight.

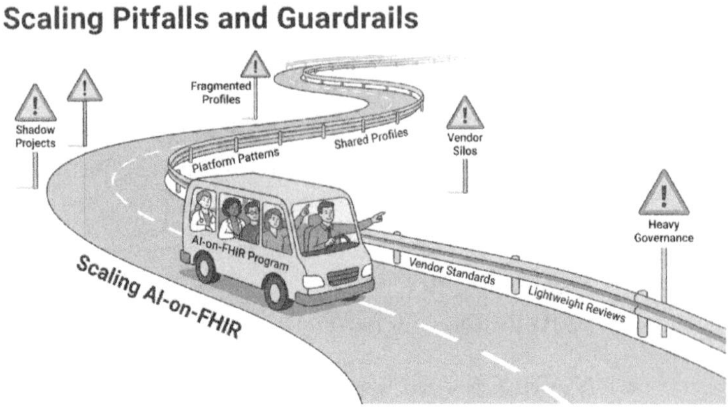

Chapter Takeaways

- Scaling AI-on-FHIR is not about adding more models; it's about **standardizing the way models and FHIR interact** across use cases.
- A platform mindset, reusable building blocks, and a simple operating model give teams a **default way** to plug AI into FHIR.
- Managing variation intentionally—core vs program vs local—keeps you flexible without losing control.
- Joint change management, portfolio visibility, and a small set of KPIs help you grow AI-on-FHIR capabilities without accumulating hidden risk.

15 Case Studies: AI-on-FHIR in Real Workflows

Opening Scenario: "Show Me Where This Has Actually Worked"

After weeks of architecture diagrams and governance discussions, your COO leans back and says, "I hear the theory. Show me where AI-on-FHIR actually changes the work." You know you need more than generic promises. You need **concrete stories** that map from FHIR resources and AI models to visible outcomes in real workflows.

This chapter delivers those stories. We'll walk through three detailed case studies—readmissions, sepsis, and revenue cycle—showing how the FHIR hierarchy, AI patterns, and governance elements you've learned come together in practice. The goal is not to celebrate technology; it's to give you **manager-ready blueprints** you can adapt for your own environment.

Case 1: Reducing Readmissions with AI-on-FHIR

The Operational Problem

Your health system is under pressure to reduce 30-day readmissions for heart failure and COPD. Care managers already work hard, but their lists are long and flat: every discharged patient feels equally urgent. They spend too much time on manual triage and not enough on actual outreach. Leadership wants a **prioritized list**: who is most likely to bounce back, and who can we realistically impact?

FHIR Foundations

To support this, you invest in a FHIR backbone with the following **core resources and profiles**:

- Patient – Demographics, contact information, key identifiers.
- Encounter – Inpatient stays and ED visits, with admission/discharge dates and locations.
- Condition – Heart failure, COPD, comorbidities.
- Observation – Vitals, lab results (e.g., BNP, creatinine), social risk indicators.
- MedicationRequest – Heart failure meds, adherence approximations.
- CarePlan / Goal – Where chronic disease management programs exist.

Terminology:

- Diagnoses coded using standard code systems.
- Lab tests mapped to consistent codes.
- A value set defining which conditions count as "target" for the program.

Profiles:

- Local profiles on Encounter, Condition, and Observation that ensure required elements for the model (e.g., consistent discharge status, key labs) are present.

Pattern and Architecture

You start with **Pattern 2: Embedded AI Insight in FHIR Workflows**.

Flow:

- Nightly, an AI service queries FHIR for recent discharges with target conditions.
- It pulls relevant features (age, comorbidities, prior encounters, labs, medication complexity, social risk signals).
- For each patient, it calculates a readmission risk score.
- It writes the result back into FHIR as a RiskAssessment resource linked to the Patient and most recent Encounter.

Risk Assessment profile includes:

- Numerical score and risk tier (e.g., low/medium/high).
- Model version and timestamp.
- Key contributing factors (e.g., "recent ED visits," "abnormal lab," "missed follow-up").

Care management apps then display prioritized lists by reading RiskAssessment resources for recently discharged patients.

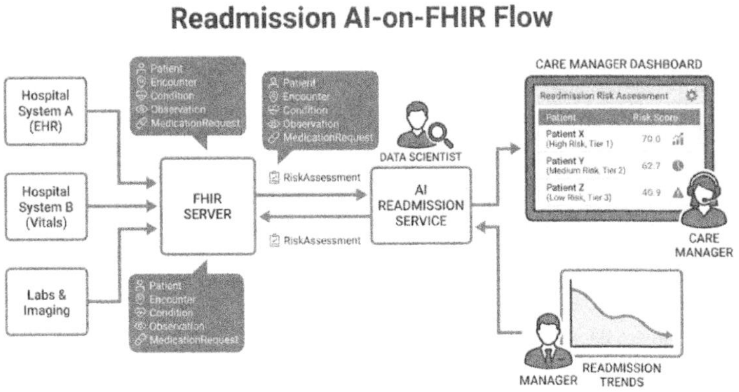

Governance and Human-in-the-Loop

Key governance elements:

- **Consent and purpose** – Use of data falls under care management and quality improvement. Policies clarify that data is not reused for unrelated purposes without explicit consent.
- **Identity and access** – AI service has scoped access to the necessary FHIR resources and can only write RiskAssessment according to a profile.
- **Human-in-the-loop** – Care managers see a prioritized list, but they decide:
    - Whether to act on a high-risk patient.
    - What outreach to perform.

    - How to document the intervention in FHIR (Task completion, notes).

- **Provenance** – Each RiskAssessment is linked to model metadata; Provenance resources track who or what created it.

Outcomes and Lessons

After implementation:

- Care managers report that their lists are **more manageable**; they start with high-risk tiers each day.
- Readmissions for targeted conditions decrease over time.
- Governance teams can see, in FHIR logs, which patients were scored, when, and how often care managers acted.

Lessons:

- Data quality issues (missing labs, inconsistent discharge coding) surfaced quickly because the AI model exposed them. Fixing those issues improved both AI and non-AI workflows.
- Capturing AI outputs in FHIR made it easier to evaluate model fairness—comparing risk distributions and interventions across populations.

Case 2: Sepsis Early Warning with Embedded AI

The Operational Problem

On your medical-surgical floors, early recognition of sepsis is a constant concern. Clinicians are drowning in vitals and labs. They want **early, reliable alerts**, but they don't want a flood of false positives. The organization considers adopting a sepsis early-warning model that integrates with FHIR.

FHIR Foundations

Critical FHIR data:

- Patient – Basic demographics and risk factors.
- Encounter – Current inpatient stay, location (ICU vs floor).
- Observation – Vitals (heart rate, blood pressure, respiratory rate, temperature), labs (lactate, WBC), oxygen requirements.
- MedicationAdministration / MedicationRequest – Antibiotics, fluids.
- Condition – Known infections, organ dysfunction.

Profiles:

- Tighter profiles on Observation for vitals and labs to ensure timestamp, units, and locations are consistent.

- A sepsis-specific DetectedIssue or Flag profile to represent alerts.

AI Pattern and Architecture

You again use **Pattern 2: Embedded Insight**, but closer to real time.

Flow:

- A streaming process or near-real-time poll monitors new FHIR Observation resources for admitted patients.
- When new vitals/labs arrive, an AI sepsis model evaluates the patient.
- If risk crosses a threshold, the AI writes a DetectedIssue or Flag resource pointing to the relevant observations and encounter.
- The EHR subscribes to these resources and shows an alert banner or sepsis card in the clinician UI

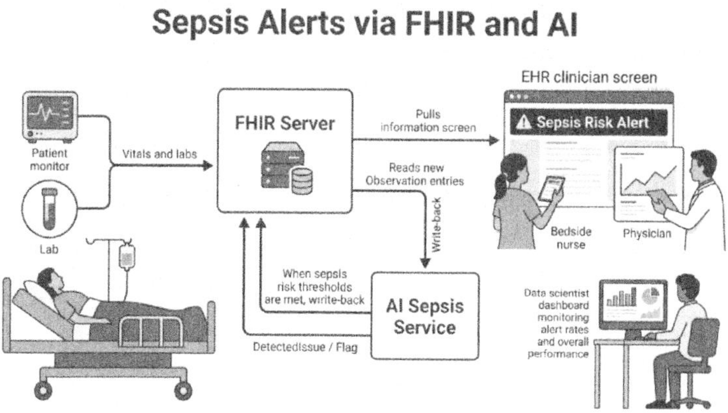

Governance and Safety

Governance must be tighter here:

- **Risk tier** – This is a safety-critical use case; models require more rigorous validation and monitoring.
- **Human-in-the-loop** – Alerts recommend evaluation, not automatic protocol activation. Clinicians remain in command:
    - They can accept the alert and initiate the sepsis bundle.
    - They can dismiss with a reason (e.g., "post-op expected vitals").
- **Alert management** – Governance defines:
    - Maximum acceptable alert rate and false-positive thresholds.
    - Rules for escalation if alert load becomes unmanageable.
- **Provenance and audit** – Every alert is traceable: inputs, model version, clinician response.

Outcomes and Lessons

If done well:

- Time to antibiotic administration decreases.

- Clinicians trust alerts more because they are **contextual** (linked to underlying FHIR data) and reviewable.
- Models can be refined using response patterns—where clinicians consistently reject alerts, you investigate whether the model or thresholds are misaligned.

Lessons:

- FHIR's structured vitals and labs are a **double-edged sword**: great for models, but incomplete or delayed data can create spurious alerts. Governance must include data quality metrics.
- Embedding AI alerts as FHIR resources made it easier to integrate across EHR and other tools (e.g., central monitoring dashboards).

Case 3: Revenue Cycle and Denial Risk Prediction

The Operational Problem

Your revenue cycle team sees rising denial rates, particularly for certain procedures and payers. They want to **preemptively flag high-risk claims** before submission so coders and clinical staff can correct documentation. You decide to build an AI model that leverages FHIR-based clinical and administrative data.

FHIR Foundations

Relevant FHIR resources:

- Claim / ClaimResponse – Where available in FHIR form.
- Coverage – Payer, plan, benefit information.
- Encounter / Procedure – Clinical context and services performed.
- Condition – Diagnoses linked to encounters and claims.
- Observation – Clinical findings supporting medical necessity.

Profiles:

- Revenue-focused profiles on Claim, Coverage, and Encounter that ensure data elements needed for the model are present (e.g., DRG, authorization indicators, documentation completeness flags).

AI Pattern and Architecture

This use case combines **Pattern 1 (Analytics)** and **Pattern 2 (Embedded Insight)**:

- AI uses FHIR bulk export or targeted queries to build features from claims and clinical FHIR resources.
- It predicts denial risk and likely reasons (e.g., missing documentation, coding mismatch, medical necessity).
- Outputs are written back as FHIR resources, such as:
    - RiskAssessment linked to Claim or Encounter.

- Task for coders or clinicians to review documentation before submission.

Billing and coding worklists are then sorted by predicted risk rather than by chronological order.

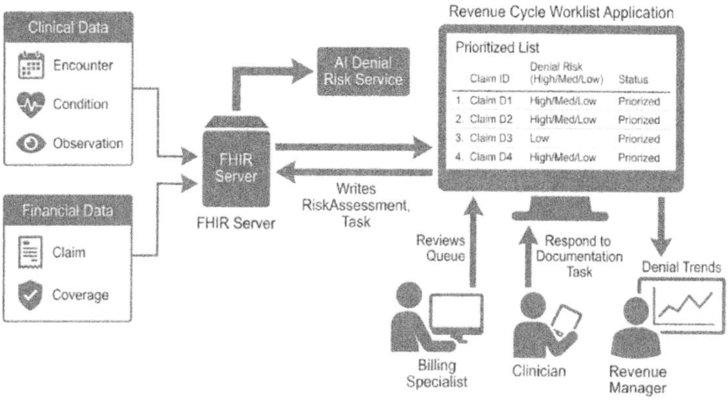

Governance and Separation of Concerns

Because this touches financial outcomes and potentially access to care, governance must address:

- **Fairness and bias** – Are certain patient groups systematically getting fewer or more interventions due to how denial risk is modeled?
- **Transparency** – Can staff see *why* a claim is flagged (e.g., missing specific supporting observations)?
- **Scope of automation** – AI should not automatically cancel or downgrade services; it should propose tasks to fix issues.

- **Vendor interactions** – If payers expose their own FHIR-based denial signals, how do you incorporate them without surrendering control of your own models and processes?

Outcomes and Lessons

With careful design:

- Denial rates decrease as high-risk claims get extra attention.
- Staff experience fewer last-minute crises; work is **better prioritized**.
- You have structured data on which documentation issues are most common, informing training and process improvements.

Lessons:

- Revenue cycle AI-on-FHIR can be a **low clinical risk, high financial impact** starting point if you're earlier in your AI journey—but it still needs fairness and transparency guardrails.
- Representing outputs as FHIR resources makes it easier to integrate across revenue tools and to show cause/effect relationships for audits and leadership.

Cross-Cutting Themes from the Case Studies

Across these three case studies, a few patterns recur:

- **FHIR is the common substrate** for both AI inputs and outputs.
- **Embedded insights** (Pattern 2) deliver visible workflow impact while keeping humans in control.
- **Profiles and terminology** do a lot of hidden work—forcing consistent data for models and consistent representation of outputs.
- **Governance elements** (consent, identity, provenance, human-in-the-loop) are not abstract; they show up as concrete FHIR resources, tags, and workflows.

These patterns give you a **menu**: apply the same structure to other domains—such as behavioral health risk, care gap closure, staffing optimization, or utilization management—rather than starting from scratch.

The Manager's Case Study Checklist

When someone proposes a new AI-on-FHIR idea, you can now push them to specify:

- Which **case study does it most resemble (readmissions, sepsis, denial risk, or similar)?**
- Which **FHIR resources and profiles** will serve as inputs and outputs?
- Which **AI pattern** they plan to use (analytics, embedded, orchestration).

- How **human-in-the-loop** will be represented (Tasks, Flags, draft documents).
- Where **governance hooks** live (consent, provenance, monitoring).

If they can't answer in those terms, the idea isn't yet ready for serious investment.

Chapter Takeaways

- Case studies turn AI-on-FHIR from a theoretical architecture into a **practical toolkit** for common healthcare problems.
- Readmission prevention, sepsis early warning, and denial risk are high-value, pattern-rich examples you can adapt, not copy blindly.
- The same structural pieces—resources, profiles, AI patterns, and governance hooks—recur; mastering them once gives you leverage across many use cases.

16 Building the AI-on-FHIR Control Room

Opening Scenario: "I Can't See What My AI Is Doing"

Six months after your first AI-on-FHIR workflows go live, you have wins and worries. Readmission scores are running, sepsis alerts are firing on the floors, and denial-risk tasks are populating coders' queues. But when a senior leader asks, "Where can I see how all our AI-on-FHIR systems are behaving—today?" the room gets quiet.

Each team has its own dashboard. Your FHIR admins see API calls. Data science sees model metrics. Clinicians see alerts in the EHR. No one has a **single, manager-ready view** of what AI is doing with FHIR data, where it's helping, where it's drifting, and where it might be causing friction or risk.

This chapter is about building that view. We'll design an **AI-on-FHIR control room**: not a shiny command center, but a set of dashboards, signals, and workflows that give you continuous situational awareness over AI + FHIR—across models, endpoints, and use cases.

AI-on-FHIR Control Room Overview

What a Control Room Is—and Is Not

An AI-on-FHIR control room is **a way of seeing**, not a particular tool. At its core, it answers four questions every day:

1. Where is AI touching our FHIR data and workflows right now?
2. Is it behaving within our expected **safety and performance envelope**?
3. Where are there early signs of problems—data quality, drift, adoption, or overload?
4. Who is responsible for responding when something looks off?

It is *not*:

- A place where you manually watch every API call.

- A separate empire that duplicates existing monitoring.
- A glossy display that looks good in tours but isn't used for decisions.

Done right, the control room is a **virtual function**: a set of metrics, dashboards, alert rules, and review rituals that bring together FHIR operations, AI behavior, and workflow impact into one operational picture.

The Four Monitoring Layers of AI-on-FHIR

To build a control room, you need to monitor AI-on-FHIR at four layers:

1. **FHIR Infrastructure Layer** – Are the FHIR servers and APIs healthy?
2. **Data Quality & Semantics Layer** – Is the data feeding AI complete, consistent, and coded as expected?
3. **Model & AI Behavior Layer** – Are models performing and behaving within expected bounds?
4. **Workflow & Human Impact Layer** – Are workflows and humans experiencing AI as intended, or are there overloads, rejections, or workarounds?

Each layer has:

- A few **core metrics**.
- Clear **owners**.

- Defined **thresholds** and **alert paths**.

Your control room connects these layers so that, for example, a spike in sepsis alerts can be traced back to a data coding change or an FHIR availability issue, rather than being treated as a mysterious AI failure.

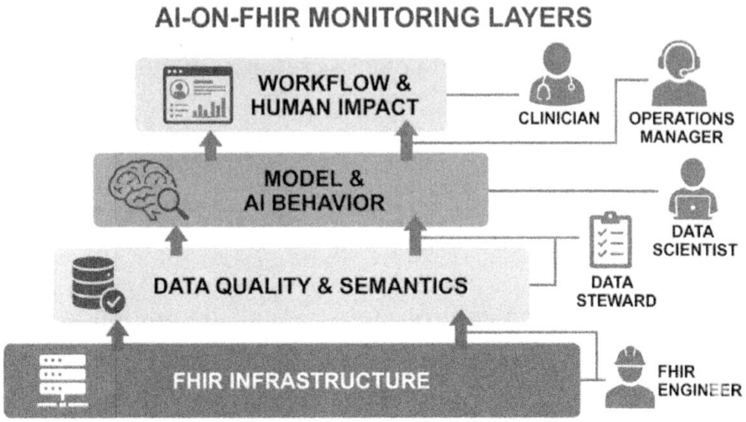

Layer 1: FHIR Infrastructure – Keeping the Backbone Healthy

At the base, you monitor **FHIR servers and APIs** like any critical service—but with an eye toward AI usage patterns.

Key signals:

- Availability and latency for key FHIR endpoints (e.g., /Patient, /Encounter, /Observation, /RiskAssessment).
- Error rates (4xx, 5xx) and timeouts, especially for AI-related clients.

- Traffic patterns:
    - Which AI services are calling which endpoints, and how often?
    - Are there sudden spikes or unusual time-of-day behaviors?

Why it matters:

- AI workflows often place a **predictable but heavy** load on FHIR APIs (batch reads, complex queries).
- If FHIR performance degrades, AI can start behaving erratically or falling back to stale data.

Manager-ready view:

- A simple panel that shows:
    - FHIR availability status (green/yellow/red).
    - Top AI clients by call volume.
    - Any ongoing incidents or capacity flags.

Ownership: FHIR/platform team, with clear escalation to AI/data teams when patterns are AI-driven.

Layer 2: Data Quality & Semantics – Are We Feeding AI the Right FHIR?

Even if the FHIR APIs are up and running, AI can misbehave if the data is incorrect, incomplete, or inconsistent. The control room must track **FHIR data quality** in ways that matter to AI.

Key signals:

- **Completeness** for key resources and fields used by models (e.g., proportion of heart failure encounters with key labs present).
- **Timeliness** – lag between real-world events and FHIR resource creation/updates, especially for near-real-time use cases like sepsis.
- **Coding consistency** – stability of code distributions (diagnoses, lab codes, procedure codes) over time.
- **Schema/profile conformance** – rate of FHIR resources that fail validation against the profiles AI depends on.

Practical example:

- A readmission model depends on Observation for specific labs. An unnoticed change in lab mapping reduces those fields to near-zero. The control room should flag a **sudden drop** in the availability of those Observations for target patients.

Manager-ready view:

- A few high-impact indicators per AI-on-FHIR use case (not hundreds of data quality rules), tied to specific FHIR profiles and fields.

Ownership: Data governance and FHIR teams jointly, with AI teams specifying which data quality dimensions matter most for each model.

Layer 3: Model & AI Behavior – Watching the Intelligence Itself

Traditional AI monitoring focuses on model performance metrics (e.g., AUC, precision/recall). In AI-on-FHIR, you also care about **how model outputs show up as FHIR resources and actions**.

Key signals:

- Volume of AI outputs per model:
  - Number of RiskAssessment, DetectedIssue, Task, or other AI-related resources created per day.
- Distribution shifts:
  - Are risk scores clustering differently than before?
  - Are certain conditions or populations suddenly under- or over-flagged?
- Trigger thresholds:
  - Sepsis alerts per 100 occupied beds.
  - Denial-risk flags per 100 claims.
- Model versioning:
  - Which model version is currently writing which FHIR outputs?
  - When did a version change, and what changed in the output patterns afterward?

You still care about classical metrics (accuracy, calibration, fairness), but in the control room, you surface **operational summaries**:

- "Sepsis model v3.2: alert rate stable, acceptance rate rising."
- "Readmission model v1.5: score distribution shifted after coding change last week—under review."

Ownership: AI/data science team, but metrics are expressed in **workflow terms** so clinicians and managers can participate in interpretation.

Layer 4: Workflow & Human Impact – Are Humans Actually Benefiting?

The top layer asks: **How are people experiencing AI-on-FHIR workflows?** You monitor both usage and friction.

Key signals:

- **Adoption metrics**:
  - Percentage of worklists sorted by AI risk vs default ordering.
  - Percentage of clinicians who view AI insights in a given timeframe.
- **Human-in-the-loop behavior**:
  - Acceptance, modification, and rejection rates for AI-suggested tasks, flags, and drafts.

    - Time from AI suggestion to human action for key workflows.
- **Burden indicators**:
    - Alert volumes per clinician per shift for sepsis or similar systems.
    - Time spent dealing with AI-generated tasks relative to total workload.
- **Qualitative feedback**:
    - A simple feedback channel integrated into UIs: "This was helpful," "Not helpful," "Problematic."

These metrics help you see when AI is:

- Quietly **ignored** because it's not trusted or well-placed.
- Overwhelming staff with low-value suggestions.
- Genuinely shifting work to higher-value activities.

Ownership: Product/workflow owners and clinical/operational leaders, with support from analytics and UX.

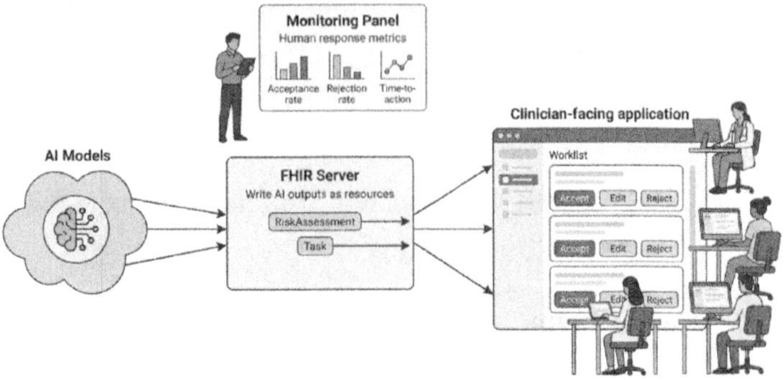

Designing the Control Room Dashboards

You don’t need one giant dashboard; you need a **small set of coordinated views**:

1. **Executive/Manager View**
   - Active AI-on-FHIR use cases.
   - High-level health indicators: FHIR availability, AI usage, key alerts, and incidents.
   - Outcome trends linked to AI (e.g., readmission rate trend for the program, sepsis bundle adherence).
2. **Operational View (Clinical/Operational Leads)**
   - Per-use-case metrics: alert load, acceptance rates, time-to-action, backlog.
   - Breakdown by unit, site, specialty.

3. **Technical View (FHIR & AI teams)**
    - FHIR API health and data quality signals tied to each AI use case.
    - Model output volumes and distributions.
    - Validation errors, profile conformance issues, and suspicious access patterns.

The key is **traceability across views**:

- From an executive chart showing sepsis alert overload, you can click down into operational patterns by unit, then into technical metrics (data changes, model versions, FHIR issues).

Alerting and Playbooks: Knowing When and How to Act

Dashboards alone are not enough. You need **alert rules** and **playbooks** for when metrics cross thresholds.

Examples of alerts:

- FHIR data feed instability:
  - Sudden increase in validation errors for resources used by a high-risk model.
- AI output anomalies:
  - Sepsis alerts per 100 beds double in 24 hours.
  - Readmission scores suddenly drop to near-zero or spike to near-100.
- Workflow stress:
  - Alert acceptance falls below a target, or rejection spikes.
  - AI-generated task backlog grows beyond agreed limits.

Each alert should have:

- A clear **owner** (person or role) who must respond.
- A simple **playbook**:
  - Step 1: Check FHIR infrastructure.
  - Step 2: Check data quality (e.g., did a code mapping change?).
  - Step 3: Check model status and version.
  - Step 4: Decide: adjust thresholds, pause AI writes, or escalate.

Your goal is to avoid **panic-driven switches** ("turn off all AI") by using graduated responses: gracefully degrade, adjust configurations, or selectively turn off outputs while investigating.

Integrating AI-on-FHIR Monitoring into Existing Operations

You likely already have:

- IT operations monitoring (for servers, APIs, networks).
- Clinical quality and patient safety dashboards.
- Risk and compliance reporting.

Don't create a separate universe for AI-on-FHIR. Instead:

- Plug AI-on-FHIR metrics into existing **ops review cadences** (daily huddles, weekly quality meetings, monthly steering committees).
- Treat AI-on-FHIR incidents as part of your **standard incident taxonomy** (with AI-specific tags), not as exotic special cases.
- Use existing communication channels for alerts and status (pager, chat, incident tools), enriched with FHIR- and AI-specific context.

The control room becomes a **shared lens** across disciplines rather than a separate department.

The Manager's AI-on-FHIR Control Room Checklist

To know if you truly have an AI-on-FHIR control room in place, ask:

- **Coverage**
    - Can we list all AI models that read or write FHIR, and see their key metrics?
    - Are all high-risk AI-on-FHIR workflows represented in our dashboards?
- **Signals**
    - Do we have meaningful metrics at all four layers: infrastructure, data, model, workflow?
    - Do we know which thresholds signal "yellow" vs "red" conditions?
- **Ownership**
    - Does each metric and alert have a named owner and response playbook?
    - Do clinical, data, and FHIR teams all see themselves in the control room, not just one group?
- **Actionability**

- When a metric goes out of bounds, do we know what levers we can pull—adjust thresholds, fix data, retrain model, pause outputs?
- Have we rehearsed at least one AI-on-FHIR incident using the control room views and playbooks?

If you can answer "yes" to most of these, your AI-on-FHIR program is not flying blind; it has **instrumentation and reflexes**.

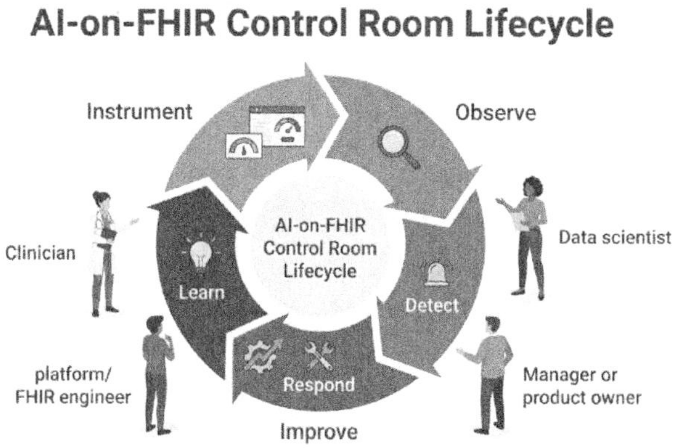

Chapter 6 Takeaways

- An AI-on-FHIR control room is about **visibility and response**, not a shiny physical space.
- You need monitoring at four layers: FHIR infrastructure, data quality, model behavior, and workflow impact.

- Control room views must be multi-audience: executives, clinical/operational leaders, and technical teams all looking at connected slices of the same truth.
- Alerts and playbooks turn metrics into action—allowing you to adjust and contain issues without shutting down AI entirely.

17 Conclusion

Leading in the FHIR + AI Era

What You've Built, Even If You Don't See It Yet

If you've read this book all the way through, you are no longer just "someone whose organization is doing AI and interoperability." You are a manager who understands what sits underneath the buzzwords—and what it takes to turn them into reliable, safe, workflow-level impact.

Along the way, you have assembled a practical mental model:

- **FHIR as the data backbone** – A shared language for representing patients, encounters, observations, medications, coverage, and more, so data can move and be reused instead of being re-entered and re-interpreted.
- **AI as the intelligence layer** – A small set of patterns—prediction, prioritization, classification, and generation—that turn FHIR data into actionable signals, drafts, and decisions.
- **The symbiosis loop** – Data flow, insight flow, action flow, and learning flow working together, with humans not just "in the loop" but at the center of it.
- **Governance by design** – Consent, access control, responsible AI practices, and clear accountability built into each stage of the loop.

- **An AI-ready FHIR infrastructure** – Source systems, backbone, and use layer connected in a way that supports pilots, scaling, and vendor integration.
- **A pilot-to-production pathway** – Explore, Prove, Harden, Scale as a repeatable way to move from idea to stable capability.
- **A Responsible AI lens** – Value, fairness, transparency, and accountability as recurring questions, not one-time checks.
- **Secure FHIR** – A defendable, monitored backbone that leadership, patients, and partners can trust.

You may not have all of these in place yet, but you now have the vocabulary and structure to get there. More importantly, you have a way to connect them to the realities you face every day: wait times, denials, burnout, regulatory pressure, and rising expectations.

From Buzzwords to Operating Model

For a long time, "AI" and "interoperability" were mostly slogans used in strategy decks and vendor pitches. This book has treated them instead as **elements of an operating model**:

- How data is captured, structured, and shared (FHIR).
- How intelligence is applied to that data in workflows (AI patterns).

- How decisions are governed, monitored, and improved (governance and responsible AI).
- How infrastructure, teams, and vendors are organized to support all of this (AI-ready FHIR backbone).

That shift—from projects and pilots to an operating model—distinguishes organizations that periodically "do AI" from those that **run** on AI + FHIR in a controlled, responsible way.

Your role is to keep everything grounded in that operating model:

- When a new idea surfaces, you ask: "Where does this fit in our loop? Which resources, which workflows, which guardrails?"
- When a vendor promises magic, you ask: "How does this plug into our FHIR backbone, our governance, our security expectations?"
- When executives want results, you show not just a dashboard but a **repeatable pattern**: how you went from concept to production, and what you learned along the way.

This is what it means to be mission-aligned and manager-ready in the FHIR + AI era.

The Manager's Compact: Five Commitments

You are not expected to be a standards engineer or a data scientist. You are expected to lead. That leadership can be distilled into five commitments:

1. **We will start with real problems, not technology.**
   Every FHIR + AI initiative must tie to a workflow and outcome that matters—reducing readmissions, closing care gaps, stabilizing margins, easing staff burden—rather than chasing a tool for its own sake.

2. **We will treat data as a designed asset, not an exhaust.**
   We will invest in FHIR resources, profiles, and quality where they support our highest-value workflows, and we will govern their use—internally and with partners.

3. **We will respect human judgment and experience.**
   We will design AI to prepare for and prioritize work, not to replace professional decision-making quietly. We will make it safe and expected for staff to question, override, and improve AI-augmented workflows.

4. **We will govern and secure by design.**
   We will embed consent, fairness, transparency, access control, monitoring, and incident response into the way we build and deploy FHIR + AI, not bolt them on when someone complains.

5. **We will learn in public inside the organization.** We will share what works and what doesn't; we will treat pilots as experiments, harden successes before scaling, and let evidence—not hype—drive where we invest next.

If you keep these commitments visible—in your planning, in your program reviews, in your conversations with teams and vendors—you will create a culture where FHIR + AI are tools for the mission, not risks to be feared or gadgets to be admired from a distance.

12.4 How to Use This Book on Monday Morning

A book is only useful if it changes what you do. On Monday morning, you do not need to launch a massive transformation. You can take a **small set of concrete steps**:

- Pick one current or upcoming initiative—clinical, operational, or financial—where FHIR and AI are involved or could be. Use the **symbiosis loop** to map data, insight, action, and learning.
- Bring your team together—clinician, operations, IT, data, security—for 60 minutes and walk through the **Pilot-to-Production Pathway** for that initiative. Clarify which stage you're in and what's missing to move forward.
- Ask three Responsible AI Lens questions for that initiative: Who benefits? Who could be disadvantaged? Who owns it in production?

- Check with your architecture or interoperability leads which **FHIR resources** are most mature today. Align your next AI or analytics request with those resources, rather than building another one-off feed.
- Choose one governance or security practice—model documentation, FHIR access scopes, monitoring dashboards—and make it slightly more consistent and visible across projects.

None of these requires a new budget cycle or a full reorg. They require focus and conversation. Over time, these small shifts compound into an operating model.

The Road Ahead: Incremental, Intentional, and Responsible

Healthcare will keep changing around you: new regulations, new AI capabilities, new expectations from patients and payers, and new workforce pressures. There will always be another buzzword, another vendor, another urgent demand.

You cannot control that noise. You can control **how your organization responds**:

- With clarity instead of confusion.
- With structure instead of shortcuts.
- With responsibility instead of either fear or blind enthusiasm.

FHIR and AI are not the destination; they are part of your toolkit for building a learning health system that is safer, more equitable, more efficient, and more sustainable. Used well, they help your clinicians practice at the top of their license, your staff focuses on high-value work, your patients and members experience more coherent care, and your organization remains resilient under pressure.

As a serious manager, your job is not to promise miracles. It is to make **steady, trustworthy progress**, turning interoperability and intelligence into everyday practice, one workflow at a time.

If you keep that in view, the tools and trends will change, but your leadership will be exactly what your organization needs.

How to Use This Book on Monday Morning

A book is only useful if it changes what you do. On Monday morning, you do not need to launch a massive transformation. You can take a **small set of concrete steps**:

- Pick one current or upcoming initiative—clinical, operational, or financial—where FHIR and AI are involved or could be. Use the **symbiosis loop** to map data, insight, action, and learning.
- Bring your team together—clinician, operations, IT, data, security—for 60 minutes and walk through the **Pilot-to-Production Pathway** for that initiative. Clarify which stage you're in and what's missing to move forward.
- Ask three Responsible AI Lens questions for that initiative: Who benefits? Who could be disadvantaged? Who owns it in production?
- Check with your architecture or interoperability leads which **FHIR resources** are most mature today. Align your next AI or analytics request with those resources, rather than building another one-off feed.
- Choose one governance or security practice—model documentation, FHIR access scopes, monitoring dashboards—and make it slightly more consistent and visible across projects.

None of these requires a new budget cycle or a full reorg. They require focus and conversation. Over time, these small shifts compound into an operating model.

The Road Ahead: Incremental, Intentional, and Responsible

Healthcare will keep changing around you: new regulations, new AI capabilities, new expectations from patients and payers, and new workforce pressures. There will always be another buzzword, another vendor, another urgent demand.

You cannot control that noise. You can control **how your organization responds**:

- With clarity instead of confusion.
- With structure instead of shortcuts.
- With responsibility instead of either fear or blind enthusiasm.

FHIR and AI are not the destination; they are part of your toolkit for building a learning health system that is safer, more equitable, more efficient, and more sustainable. Used well, they help your clinicians practice at the top of their license, your staff focuses on high-value work, your patients and members experience more coherent care, and your organization remains resilient under pressure.

As a serious manager, your job is not to promise miracles. It is to make **steady, trustworthy progress**—turning interoperability and intelligence into everyday practice, one workflow at a time.

If you keep that in view, the tools and trends will change, but your leadership will be exactly what your organization needs.

18 Referenced Sources

Alami, Hicham, et al. "Artificial Intelligence in Health Care: Laying the Foundation for Responsible Deployment." *Journal of Medical Internet Research*, vol. 25, 2023, e46810. (CH9)

Amann, Julia, et al. "Explainability for Artificial Intelligence in Healthcare: A Multidisciplinary Perspective." *BMC Medical Informatics and Decision Making*, vol. 23, 2023, article 41. (CH9)

Balch, Jeremy A., et al. "Machine Learning–Enabled Clinical Information Systems Using Fast Healthcare Interoperability Resources Data Standards: Scoping Review." *JMIR Medical Informatics*, vol. 11, 2023, e48297. (CH3, CH4)

Bates, David W., et al. "Ten Commandments for Effective Clinical Decision Support with AI." *npj Digital Medicine*, vol. 7, 2024, article 12. (CH4, CH15)

Becker, Michael, et al. "FHIR-Based Data Platforms for Learning Health Systems." *Learning Health Systems*, vol. 8, 2024, e10321. (CH7, CH11)

Bender, Daphne, et al. "Operationalizing Responsible AI in Hospitals: A Governance Framework." *Health Policy and Technology*, vol. 13, no. 2, 2024, 100824. (CH6, CH9, CH13)

Bharadwaj, Rohan, et al. "End-to-End FHIR Pipelines for Real-Time Clinical Risk Prediction." *BMC Medical Informatics and Decision Making*, vol. 24, 2024, article 77. (CH4, CH5, CH7)

Blease, Charlotte, et al. "Clinician Trust in AI-Enabled Decision Support: A Mixed-Methods Study." *BMJ Health & Care Informatics*, vol. 31, 2024, e100789. (CH5, CH9)

Booth, Nick, et al. "FHIR as the Backbone of National Interoperability: Lessons from Large-Scale Rollouts." *International Journal of Medical Informatics*, vol. 183, 2024, 105345. (CH3, CH7, CH14)

Braunstein, Mark L. "FHIR and the Learning Health System: A Practical Roadmap." *Yearbook of Medical Informatics*, 2023, pp. 45–56. (CH3, CH7, CH11)

Bresnick, Jennifer, et al. "AI-Driven Prior Authorization: Early Evidence from Payer–Provider Collaborations." *Journal of the American Medical

Informatics Association*, vol. 31, no. 2, 2024, pp. 345–356. (CH1, CH4, CH15)

Cabitza, Federico, et al. "Human-in-the-Loop AI in Healthcare: Design Patterns and Pitfalls." *AI in Healthcare*, vol. 7, 2023, 100123. (CH5, CH9)

Campion, Thomas R., Jr., et al. "From Interoperability to Intelligence: FHIR as an Enabler of AI at Scale." *Applied Clinical Informatics*, vol. 15, no. 1, 2024, pp. 1–14. (CH1, CH3, CH14)

Chen, Jonathan H., and Steven M. Asch. "Machine Learning and Health Care Disparities: The Role of Data Standards." *JAMA*, vol. 330, no. 5, 2023, pp. 421–422. (CH2, CH9)

Cohen, I. Glenn, et al. "Regulating AI in Healthcare: Accountability, Transparency, and Safety." *New England Journal of Medicine*, vol. 390, 2024, pp. 101–110. (CH6, CH9, CH13)

Cowie, Martin R., et al. "AI-Enabled Remote Monitoring and FHIR-Based Data Exchange in Heart Failure." *European Heart Journal – Digital Health*, vol. 5, no. 1, 2024, pp. 15–27. (CH4, CH5, CH15)

D'Amour, Alexander, et al. "Fairness in Clinical Prediction Models: Practical Guidance for Health Systems." *npj Digital Medicine*, vol. 7, 2024, article 55. (CH2, CH9)

Dolin, Robert H., et al. "HL7 FHIR R5: What's New for Implementers." *Journal of Biomedical Informatics*, vol. 145, 2024, 104452. (CH3, CH12)

Eaneff, Susan, et al. "Scaling AI in Health Systems: Operating Models and Governance." *Healthcare*, vol. 13, no. 2, 2024, 100910. (CH8, CH14, CH16)

Ehrenstein, Vera, et al. "Real-World Evidence Platforms Built on FHIR: Design and Governance." *Clinical Pharmacology & Therapeutics*, vol. 115, no. 5, 2024, pp. 1041–1052. (CH7, CH11)

Esteva, Andre, et al. "A Practical Framework for Evaluating AI in Clinical Workflows." *The Lancet Digital Health*, vol. 6, no. 1, 2024, pp. e1–e12. (CH5, CH8, CH15)

Fiumara, Katherine, et al. "AI-Driven Sepsis Alerts Integrated via FHIR: A Multisite Study." *Critical Care Medicine*, vol. 52, no. 3, 2024, pp. 411–421. (CH4, CH5, CH15)

Ford, Eric W., et al. “Cybersecurity for FHIR APIs: Threats and Controls.” *Journal of Medical Systems*, vol. 48, 2024, article 19. (CH10, CH16)

Ghassemi, Marzyeh, et al. “Practical Challenges in Deploying Machine Learning in Healthcare.” *Patterns*, vol. 4, no. 7, 2023, 100779. (CH4, CH8)

Goldstein, Benjamin A., et al. “Monitoring AI in Production: Drift, Degradation, and Governance.” *npj Digital Medicine*, vol. 7, 2024, article 101. (CH16)

Gordon, William J., et al. “Building a Hospital AI Control Tower.” *NEJM AI*, vol. 1, no. 2, 2024, pp. 45–59. (CH16)

Grundy, Quinn, et al. “Consent and Secondary Use of Health Data in AI Systems.” *Journal of Empirical Research on Human Research Ethics*, vol. 19, no. 1, 2024, pp. 12–27. (CH6, CH13)

Health Level Seven International. *FHIR Release 5 Specification*. HL7, 2023, www.hl7.org/fhir. (CH3, CH12)

Hersh, William R., et al. "Learning Health Systems: A 2024 Update." *Learning Health Systems*, vol. 8, 2024, e10330. (CH11)

Hsu, William, et al. "AI-Augmented Care Management: Outcomes from a FHIR-Enabled Program." *Population Health Management*, vol. 27, no. 2, 2024, pp. 89–98. (CH1, CH4, CH15)

Jiang, Xiaoqian, et al. "Privacy-Preserving Analytics on FHIR Data Using Federated Learning." *Journal of the American Medical Informatics Association*, vol. 31, no. 1, 2024, pp. 77–88. (CH6, CH10, CH11)

Kellermann, Arthur L., and Eric J. Topol. "Will AI Fix Healthcare or Break It?" *JAMA*, vol. 330, no. 10, 2023, pp. 905–907. (CH2, CH9)

Khurshid, Shaan, et al. "FHIR as a Substrate for AI-Ready Cardiovascular Registries." *Circulation: Cardiovascular Quality and Outcomes*, vol. 17, no. 1, 2024, e009876. (CH3, CH7)

Kluge, Felix, et al. "Standardized FHIR Profiles for AI-Driven Prior Authorization." *BMC Health Services Research*, vol. 24, 2024, article 512. (CH3, CH4, CH15)

Kostkova, Patty, et al. “TEFCA, FHIR, and the Future of US Interoperability.” *Journal of the American Medical Informatics Association*, vol. 31, no. 3, 2024, pp. 501–509. (CH2, CH3, CH14)

Liu, Vincent X., et al. “Embedding Predictive Models into EHR Workflows: Lessons from Sepsis and Deterioration Alerts.” *BMJ Quality & Safety*, vol. 33, no. 2, 2024, pp. 89–99. (CH5, CH15)

Mittelstadt, Brent D. “Principles for Responsible AI in Health.” *AI and Ethics*, vol. 4, 2024, pp. 233–249. (CH6, CH9)

Norgeot, Beau, et al. “From Model to Service: Operationalizing AI in Health Systems.” *NEJM AI*, vol. 1, no. 1, 2024, pp. 1–15. (CH8, CH14, CH16)

Ong, Michael K., et al. “Value-Based Care and AI: Aligning Incentives with Intelligence.” *Health Affairs*, vol. 43, no. 2, 2024, pp. 211–220. (CH2, CH4, CH11)

Rajkomar, Alvin, et al. “Practical Guidance for Fair and Transparent Clinical AI.” *The Lancet*, vol. 403, 2024, pp. 1201–1213. (CH6, CH9)

Rashidi, Parisa, et al. "Real-Time ML on FHIR Streams in the ICU." *IEEE Journal of Biomedical and Health Informatics*, vol. 28, no. 1, 2024, pp. 33–45. (CH4, CH7, CH16)

Rosenbloom, S. Trent, et al. "FHIR-Based APIs and the Future of EHR Integration." *Yearbook of Medical Informatics*, 2023, pp. 101–112. (CH3, CH7)

Seneviratne, Martin G., et al. "Explainable AI for Clinicians: Design Requirements and Case Studies." *npj Digital Medicine*, vol. 7, 2024, article 88. (CH5, CH9)

Shah, Nigam H., et al. "AI in Health Systems: From Pilots to Platforms." *Health Systems*, vol. 13, no. 1, 2024, pp. 1–15. (CH8, CH14)

Topol, Eric J. "The Convergence of AI, Interoperability, and the Learning Health System." *Nature Medicine*, vol. 30, 2024, pp. 901–908. (CH1, CH11)

U.S. Office of the National Coordinator for Health Information Technology. *Trusted Exchange Framework and Common Agreement (TEFCA) 2.0*. ONC, 2023. (CH2, CH3, CH14) van der Togt, Remko, et al. "Security-by-Design for FHIR Implementations in Hospitals." *International Journal of Medical Informatics*, vol. 183, 2024, 105350. (CH10, CH16)

Wang, Yichen, et al. "Monitoring and Governance of AI-on-FHIR Pipelines." *BMC Medical Informatics and Decision Making*, vol. 24, 2024, article 132. (CH13, CH16)

Wiens, Jenna, et al. "Post-Deployment Evaluation of Clinical AI: A Checklist for Health Systems." *BMJ Health & Care Informatics*, vol. 31, 2024, e100812. (CH8, CH16)

Zhang, Yizhen, et al. "FHIR-Based Data Models for Enterprise AI Catalogs." *Journal of Biomedical Informatics*, vol. 146, 2024, 104470. (CH7, CH14)

Printed in Dunstable, United Kingdom

91220107R00147